GENETIC DIVERSITY IN PLANT SPECIES – CHARACTERIZATION AND CONSERVATION

Edited by **Mohamed A. El-Esawi**

Genetic Diversity in Plant Species – Characterization and Conservation
http://dx.doi.org/10.5772/intechopen.75369
Edited by Mohamed A. El-Esawi

Contributors

Devendra Jain, Masatoshi Funabashi, Andrés Cortés, Paola Hurtado, Matthew W. Blair, Maria Isabel Chacón, Amaury Martín Arzate-Fernández, Rosa Laura Heredia-Bobadilla, Guadalupe Gutierrez-Gonzalez, Sergio Franco-Maass, Kamile ULUKAPI, Mohamed A. El-Esawi

Notice

First published in London, United Kingdom, 2019 by IntechOpen
IntechOpen is the global imprint of INTECHOPEN LIMITED, registered in England and Wales, registration number: 11086078, The Shard, 25th floor, 32 London Bridge Street
London, SE19SG – United Kingdom
Printed in Croatia

British Library Cataloguing-in-Publication Data
A catalogue record for this book is available from the British Library

Additional hard copies can be obtained from orders@intechopen.com

Genetic Diversity in Plant Species – Characterization and Conservation, Edited by Mohamed A. El-Esawi
p. cm.
Print ISBN 978-1-83880-409-1
Online ISBN 978-1-83880-410-7
eBook (PDF) ISBN 978-1-83880-651-4

Meet the editor

Dr. Mohamed Ahmed El-Esawi is currently a visiting research fellow at the University of Cambridge in the United Kingdom and an associate professor of molecular genetics at the Botany Department of Tanta University in Egypt. Dr. El-Esawi received his BSc and MSc from Tanta University, and his PhD degree in Plant Genetics and Molecular Biology from Dublin Institute of Technology, Technological University Dublin, in Ireland. Afterward, Dr. El-Esawi joined the University of Warwick in the United Kingdom, the University of Sorbonne (Paris VI) in France, and the University of Leuven (KU Leuven) in Belgium as a visiting research fellow. His research focuses on plant genetics, genomics, molecular biology, molecular physiology, developmental biology, plant-microbe interaction, and bioinformatics. He has authored several international journal articles and book chapters and has participated in more than 60 conferences and workshops worldwide. Dr. El-Esawi is currently involved in several research projects on biological sciences.

Contents

Preface

Genetic diversity is the raw material that helps plant species face a wide range of daily global changes. It also represents the number of alternations in the genetic makeup of populations and species that take place under various evolutionary mechanisms. Evaluation of the genetic diversity and population structure of plant germplasm enhances breeding and management practices to develop improved varieties. Various genetic tools have been successfully used to study the genetic diversity of plant species, including morphological, cytological, biochemical, and molecular markers. This book discusses the fundamental advances related to assessment, utilization, and conservation of genetic diversity in plant germplasm. The book also discusses approaches related to environmental stresses. Moreover, it sheds new light on the current research trends and future research directions related to plant genetic diversity studies. This book will provoke the interest of various readers, researchers, and scientists, who will find this information useful for the advancement of their plant genetic diversity research work.

The book includes six chapters. The first introductory chapter "Assessment and Conservation of Genetic Diversity in Plant Species" presents an introduction to plant genetic diversity and the markers used in its characterization. The second chapter "Morphological, Biochemical, and Molecular Characterization of Orange Fleshed Sweet Potato (*Ipomoea batatas* [L.] Lam) Germplasms" evaluates the genetic diversity of sweet potato germplasm based on different morphological, biochemical, and molecular markers. The third chapter "Bean Genome Diversity Reveals the Genomic Consequences of Speciation, Adaptation, and Domestication" reviews whether genomic islands of speciation are repeatedly more prone to harbor within-species differentiation due to genomic features, such as suppressed recombination, smaller effective population size, and increased drift, across repeated hierarchically nested levels of divergence. The fourth chapter "Induced Mutation: Creating Genetic Diversity in Plants" discusses the approaches related to induced mutations and genetic diversity creation in plants. The fifth chapter "Water and Ecosystem Cycles Mediated by Plant Genetic Resources for Food and Agriculture" overviews the water and ecosystem cycles mediated by the ecosystem functions of naturally occurring plant communities and discusses possibilities for the transformation of agriculture into sustainable modality with primary importance given to the recovery of water cycles. The sixth chapter "Genetic Variability of Mountain Pine (*Pinus hartwegii* Lindl) in the Protection of Flora and Fauna Area Nevado de Toluca" studies the genetic diversity of mountain pine under the attack of bark beetles and dwarf mistletoes, to generate information that could be used to improve strategies of conservation of these forests.

The book editor would like to thank Ms. Jasna Bozic, Author Service Manager, for her wholehearted cooperation in the publication of this book.

Mohamed Ahmed El-Esawi, PhD
Sainsbury Laboratory
University of Cambridge
Cambridge, United Kingdom

Botany Department
Faculty of Science
Tanta University, Egypt

Introductory Chapter: Assessment and Conservation of Genetic Diversity in Plant Species

Mohamed A. El-Esawi

Additional information is available at the end of the chapter

http://dx.doi.org/10.5772/intechopen.86060

1. Genetic diversity and its assessment

Genetic diversity is the raw material that helps plant species face a wide range of daily global changes [1, 2]. It also represents the number of alternations in the genetic makeup of populations and species, which take place under various evolutionary mechanisms such as genetic drift that involves random matings of individuals within the same population, changing their allele frequencies, and founder effect that causes individuals with new genetic information leading to a new population developed from a larger one [1–3]. Genetic composition of individuals varies from one population to another due to systematic differences that emerge among individuals from different places promoting their survival and reproduction. Genetic diversity becomes more potent and quick when gene flow among populations is little such as restricted dispersal of seeds or pollens [2, 3]. Plant breeders are able to develop large amount of new productive crops that are of improved tolerance to a variety of diseases and pests as well as its enhanced ability to stand against a changing world. This depends on the volume and range of genetic variation among individuals within the same species, which allows for designing sampling programs [2, 4–6]. Accordingly, geneticists have focused on evaluating genetic differences within populations using morphological, cytological, biochemical, and molecular markers to identify the characteristics of domestication, propagation, and breeding techniques as well as conservation of plant genetic materials [2]. This work addresses various approaches related to genetic diversity in plants.

2. Biochemical markers

2.1. Storage proteins

One important output of post-transcription and translation processes is the protein, which represents the genetic DNA as well as the structural and enzymatic material of an organism's cells [2]. Genetic divergence is found to have impacts on proteins. For example, seed proteins that are of essential roles in species are tolerant to environmental occurrences. Recent approaches have employed several technologies to identify the characteristics of various plant cultivars and genotypes [2]. One quickly and precisely technology utilized for such purpose is electrophoresis. The genetic divergence was assessed within *Lathyrus sativus* through accurate electrophoretic analysis of its seed storage proteins [2, 7], suggesting its fundamental role in evaluating the relationship between taxonomy and genetics at and below the species level [2].

2.2. Isozyme markers

Isozymes, also called isoenzymes, are proteins exhibiting same catalytic and quantitative function as the enzyme but, meanwhile, they differ in their molecular forms [2]. Various alleles within a single locus are encoded by structural genes to give the allozyme, an allelic variant of the enzyme. Biological analysis of isozymes reveals significant importance, including the assessment of genetic divergence as well as phylogenetic and taxonomic relationships [2]. It also helps to investigate population genetics and developmental biology as well as to conserve the genetic resources of plants [8, 9].

3. Molecular markers

Molecular markers such as restricted fragment length polymorphism (RFLP), random amplified polymorphic DNA (RAPD), amplified fragment length polymorphism (AFLP), microsatellites (SSRs), and single-nucleotide polymorphisms (SNPs) are considered as effective and efficient tools for evaluating the genetic divergence within and among plant species [10].

3.1. Restriction fragment length polymorphism (RFLP)

RFLP is a dominant marker that involves breakages of DNA bonds at specific nucleotides by the action of enzymes called restriction endonucleases [8, 10], followed by size fractionation of digested fragments by electrophoresis technique, suggesting for these markers a key role in setting up genetic mapping as well as evaluating genetic diversity and phylogenetic relationships. RFLP markers could be a possible option to identify the characteristics of individuals, to estimate segregational analysis of progenitors as well as to evaluate genetic variation and phylogenetic relationships in the germplasms of lettuce plants, but these techniques proven to be expensive, technically complicated, and away from optimal performance [10]. Therefore, research is needed to develop low cost and more efficient molecular genetic technologies to inhibit or even limit the technical obstacles related to RFLP technique and to study the properties of genetic difference within plants [10].

3.2. Random amplified polymorphic DNA (RAPD)

RAPD, a PCR-based technique or (AP-PCR), involves the use of arbitrary short primers [8, 10] in a PCR reaction to amplify random sequences from DNA template. RAPD has proven to be a multipurpose technique utilized for constructing genetic map [11, 12], utilization in breeding approaches [13], identification of resistant genes and hybrid origin [14], and for assessing plant genetic variance by characterizing differences between populations of the similar germplasm resources [11, 12].

3.3. Amplified fragment length polymorphism (AFLP)

AFLP is a highly reproducible marker that utilizes PCR technique to amplify DNA fragments [8, 10]. It is a DNA fingerprinting-based technique that includes digestion of DNA into small fragments with the help of restriction enzymes. These fragments are ligated by adaptors complementary to their restriction sites followed by selective amplification of the newly formed subset by PCR technique [10]. Autoradiographic and fluorescence technologies are then utilized to visualize the amplified fragments on polyacrylamide gels. AFLP is proven its successful participation in identifying the characteristics of genetic diversity and relationships of plant species [8, 10, 15–20].

3.4. Microsatellites (SSRs)

Microsatellites are repetitive sequences of nitrogen bases within DNA. These repeats may be mono-, di-, tri-, tetra-, or penta-nucleotides found in eukaryotic nuclear genome [8, 10]. Microsatellites are genetically different. Therefore, they are employed in estimating genetic divergence and recognizing the relationships between plant genotypes [10, 21].

In conclusion, morphological, cytological, biochemical, and molecular markers proved useful in assessing genetic diversity levels in different plant species [22–30].

Author details

Mohamed A. El-Esawi[1,2]*

*Address all correspondence to: mohamed.elesawi@science.tanta.edu.eg

1 The Sainsbury Laboratory, University of Cambridge, Cambridge, United Kingdom

2 Botany Department, Faculty of Science, Tanta University, Tanta, Egypt

References

[1] El-Esawi MA. Assessing the genetic diversity, phylogenetic relationships, and disease resistance genes in Irish *Brassica oleracea* species [PhD thesis]. Ireland: Dublin Institute of Technology, Technological University Dublin; 2012

[2] El-Esawi MA. Taxonomic relationships and biochemical genetic characterization of *Brassica* resources: Towards a recent platform for germplasm improvement and utilization. Annual Research & Review in Biology. 2015;**8**(4):1-11

[3] Falk DA, Knapp E, Guerrant EO. Introduction to Restoration Genetics. USA: Society for Ecological Restoration; 2001

[4] El-Esawi M, Bourke P, Germaine K, Malone R. Assessment of morphological variation in Irish *Brassica oleracea* species. Journal of Agricultural Science. 2012;**4**:20-34

[5] El-Esawi MA, Sammour R. Karyological and phylogenetic studies in the genus *Lactuca* L. (Asteraceae). Cytologia. 2014;**79**:269-275

[6] Yu J, Mosjidis J, Klingler K, Woods F. Isozyme diversity in North American cultivated red clover. Crop Science. 2001;**41**:1625-1628

[7] Sammour R, Mustafa A, Badr S, Tahr W. Genetic variations in accessions of *Lathyrus sativus* L. Acta Botanica Croatica. 2007;**66**:1-13

[8] Dziechciarková M, Lebeda A, Doležalová I, Astley D. Characterization of *Lactuca* spp. germplasm by protein and molecular markers—A review. Plant, Soil and Environment. 2004;**50**:47-58

[9] Kumar P, Gupta VK, Misra AK, Modi DR, Pandey BK. Potential of molecular markers in plant biotechnology. Plant Omics Journal. 2009;**2**:141-162

[10] El-Esawi MA. Molecular genetic markers for assessing the genetic variation and relationships in *Lactuca* germplasm. Annual Research & Review in Biology. 2015;**8**(5):1-13

[11] Waycott W, Fort S. Differentiation of nearly identical germplasm accessions by a combination of molecular and morphological analyses. Genome. 1994;**85**:239-244

[12] Kesseli R, Paran I, Michelmore R. Analysis of a detailed genetic linkage map of *Lactuca sativa* (lettuce) constructed from RFLP and RAPD markers. Genetics. 1994;**136**:1435-1446

[13] Montesclaros L, Nicol N, Ubalijoro E, Leclerc-Potvin C, Ganivet L, Laliberté J. Response to potyvirus infection and genetic mapping of resistance loci to potyvirus infection in *Lactuca*. Theoretical and Applied Genetics. 1997;**94**:941-946

[14] Friesen N, Fritsch R, Bachmann K. Hybrid origin of some ornamentals of *Allium* subgenus *Melanocrommyum* verified with GISH and RAPD. Theoretical and Applied Genetics. 1997;**95**:1229-1238

[15] Koopman W, Zevenbergen M, Ronald G, van den Berg R. Species relationship in *Lactuca* S.L. (Lactuceae, *Asteraceae*) inferred from AFLP fingerprints. American Journal of Botany. 2001;**88**:1881-1887

[16] Hill M, Witsenboer H, Zabeau M, Vos P, Kesseli R, Michelmore R. PCR-based fingerprinting using AFLPs as tool for studying genetic relationship in *Lactuca* spp. Theoretical and Applied Genetics. 1996;**93**:1202-1210

[17] Johnson WC, Jackson LE, Ochoa O, van Wijk R, Peleman J, St Clair DA, et al. Lettuce, a shallow-rooted crop, and *Lactuca serriola*, its wild progenitor, differ at QTL determining

root architecture and deep soil water exploitation. Theoretical and Applied Genetics. 2000;**101**:1066-1073

[18] Jeuken M, van Wijk R, Peleman J, Lindhout P. An integrated interspecific AFLP map of lettuce (*Lactuca*) based on two *L. sativa* × *L. saligna* F2 populations. Theoretical and Applied Genetics. 2001;**103**:638-647

[19] Jeuken M, Lindhout P. *Lactuca saligna*, a non-host for lettuce downy mildew (*Bremia lactucae*), harbors a new race-specific *Dm* gene and three QTLs for resistance. Theoretical and Applied Genetics. 2002;**105**:384-391

[20] Kuang H, van Eck HJ, Sicard D, Michelmore R, Nevo E. Evolution and genetic population structure of prickly lettuce (*Lactuca serriola*) and its RGC2 resistance gene cluster. Genetics. 2008;**178**(3):1547-1558

[21] Riar D, Rustgi S, Burke I, Gill K, Yenish J. EST-SSR development from 5 *Lactuca* species and their use in studying genetic diversity among *L. serriola* biotypes. The Journal of Heredity. 2011;**102**:17-28

[22] El-Esawi MA, Germaine K, Bourke P, Malone R. Genetic diversity and population structure of *Brassica oleracea* germplasm in Ireland using SSR markers. Comptes Rendus Biologies. 2016;**339**:133-140

[23] El-Esawi MA, Germaine K, Bourke P, Malone R. AFLP analysis of genetic diversity and phylogenetic relationships of *Brassica oleracea* in Ireland. Comptes Rendus Biologies. 2016;**339**:163-170

[24] El-Esawi MA, Mustafa A, Badr S, Sammour R. Isozyme analysis of genetic variability and population structure of *Lactuca* L. germplasm. Biochemical Systematics and Ecology. 2017;**70**:73-79

[25] El-Esawi MA. Genetic diversity and evolution of *Brassica* genetic resources: From morphology to novel genomic technologies—A review. Plant Genetic Resources. 2017;**15**:388-399

[26] El-Esawi MA. SSR analysis of genetic diversity and structure of the germplasm of faba bean (*Vicia faba* L.). Comptes Rendus Biologies. 2017;**340**:474-480

[27] El-Esawi MA, Al-Ghamdi AA, Ali HM, Alayafi AA, Witczak J, Ahmad M. Analysis of genetic variation and enhancement of salt tolerance in French pea (*Pisum sativum* L.). International Journal of Molecular Sciences. 2018;**19**:2433

[28] El-Esawi MA, Alaraidh IA, Alsahli AA, Ali HM, Alayafi AA, Witczak J, et al. Genetic variation and alleviation of salinity stress in barley (*Hordeum vulgare* L.). Molecules. 2018;**23**:2488

[29] El-Esawi MA. Micropropagation technology and its applications for crop improvement. In: Anis M, Ahmad N, editors. Plant Tissue Culture: Propagation, Conservation and Crop Improvement. Singapore: Springer; 2016. pp. 523-545

[30] El-Esawi MA. Nonzygotic embryogenesis for plant development. In: Anis M, Ahmad N, editors. Plant Tissue Culture: Propagation, Conservation and Crop Improvement. Singapore: Springer; 2016. pp. 583-598

Chapter 2

Morphological, Biochemical, and Molecular Characterization of Orange-Fleshed Sweet Potato (*Ipomoea batatas* [L.] Lam) Germplasms

Anubhuti Sharma, Devendra Jain,
Sunil Kumar Khandelwal, Ravish Chaudhary,
Kapil Dev Ameta and Abhijeet Singh

Additional information is available at the end of the chapter

http://dx.doi.org/10.5772/intechopen.82597

Abstract

The sweet potato is considered as an excellent source of β-carotene and anthocyanins and has a considerable value in the functional food market. In this report, 21 sweet potato (*Ipomoea batatas* [L.] Lam) germplasms were evaluated for genetic diversity using morphological and biochemical and molecular markers. Ten morphological traits were studied, and the mean squares due to germplasm were highly significant for storage root number per plant, individual root weight, storage root (fresh) per plant, storage root (dry) per plant, storage root yield, and storage root length. UPGMA cluster analysis based on morphological traits separated the germplasm into three groups. The similarity coefficient ranged from 0.00 to 0.50 with an average of 0.176. Biochemical analysis, viz. total phenol and antioxidant, was performed to find out superior genotype at biochemical level under given conditions. Maximum total phenol was observed in the genotype "V-12" (1.39 mg), whereas maximum total antioxidant was observed in "Samrat" (0.30 mg). RAPD analysis was carried out, and out of 15 RAPD primers, 10 primers produced 96 reproducible and polymorphic bands. UPGMA cluster analysis based on RAPD data also separated the genotypes into three clusters. The results of the present study can be used for sweet potato crop improvement through molecular breeding and marker-assisted selection for desired traits in future.

Keywords: sweet potato, morphological marker, biochemical, RAPD, UPGMA

1. Introduction

The sweet potato (*Ipomoea batatas* [L.] Lam) is a dicotyledonous plant that belongs to the family Convolvulaceae and plays a critical role in food security after potato and cassava. Its large, starchy, sweet-tasting, tuberous roots are most important root crop worldwide [1]. The young leaves and shoots are sometimes eaten as greens. The sweet potato is native to the tropical regions in America, and its cultivation area covers around 135,000 hectares with an estimated annual production of 1,639,000 MT in India [2]. The roots are used as a source of carbohydrate and dietary fiber. Dietary fiber has the potential to reduce the incidence of a variety of diseases in man, including colon cancer, diabetes, heart diseases, and digestive disturbances [3]. Orange, white, and creamy flesh sweet potato is most commonly grown and eaten. In orange- and yellow-fleshed sweet potato, color is due to the presence of carotenoids of which β-carotene is most abundant. To increase the yield and quality of the sweet potato, it is important to study on molecular and biochemical variation in sweet potato genotypes, but unfortunately, negligible work has been done on sweet potato in spite of fact that lot of variability exists in sweet potato for physiological and biochemical characters, which can be utilized for improving tuber yield coupled with high nutritive value [4].

Food fortification, dietary diversification, and vitamin A supplementation are the recommended strategies to control vitamin A deficiency. Physiological and biochemical factors determine the storage quality of any crop. As the tuber forms major proportion of total dry weight of plant, productivity is largely governed by the process of tuberization and photosynthetic efficiency of the leaf canopy in support of the storage root sink. Both the processes are being controlled by environmental factors [5]. The flesh color of the root varies from various shades of white, cream, yellow to dark-orange depending upon the carotenoid content. In the orange-fleshed sweet potato, the major carotenoid present is β-carotene. Carotenoids have been linked with the enhancement of immune system and decreased risk of degenerative diseases such as cardiovascular problems, age-related macular degeneration, and cataract formation [6]. Interest in carotenoids has increased due to their possible health benefits as carotenoids are often associated with health preventive effort and reduced risk of aged related macular degeneration, anticancerogenic activity, antioxidant capacity, antiulcer activity, and also reduced risk of cardiovascular disorders [7]. Both sweet potato roots and tops are nutritious foods, which could be used to advantage in combating nutritional deficiencies in parts of the developing world if means could be found to overcome resistance to their increased consumption. The roots are used primarily as human food and are eaten boiled, steamed, fried, or baked. Raw leaves and tuber tips are also excellent sources of ascorbic acid and some of the vitamin B, especially riboflavin that is deficient in many Asian diets. However, high percentages of water soluble vitamin are lost on cooking [8]. Recent studies associated with the consumption of carotenoid-rich food showed the decrease of the incidence of certain cancers in human beings [9].

Assessment of genetic diversity at the molecular level is more meaningful than at the phenotypic level as the later involves data on morphological traits, which are environmental dependent. Different molecular marker systems have been successfully employed to assess the genomic stability of regenerated plants regardless of the presence or absence of obvious phenotypic alterations earlier. Molecular markers have been proved to be valuable tools in the characterization and evaluation of genetic diversity within and between species and population. It has been deserved that different markers might reveal different classes of variation [10, 11]. RAPD markers offer many advantages such as higher frequency of polymorphism, rapidity [12], technical simplicity, requirement of a few nanograms of DNA, no requirement of prior information of any DNA sequence, and feasibility of automation [13]. Hence, the present study is the attempt to determine genetic diversity among sweet potato (*Ipomoea batatas* [L.] Lam) genotypes using morphological, biochemical, and molecular markers.

2. Materials and methods

2.1. Plant material

Morphological and Molecular Characterization of twenty one diverse germplasms of sweet potato (*Ipomoea batatas*) were reported in this research. The experimental materials comprised of 21 diverse germplasms of sweet potato were sown in randomized block design replicated thrice. Each entry was planted in 3 × 2.4 m^2, keeping row to row and plant to plant distance of 60 and 30 cm, respectively. The recommended packages of practices were followed to raise a healthy crop.

2.2. Morphological analysis

The genotypes were evaluated in randomized block design with three replications. Observations on various morphological characters like storage root number per plant, individual root weight (g), storage root (fresh) per plant, storage root dry yield per plant (g), storage root yield per plot (kg), and storage root length (cm) were recorded from five selected plants from each replication, which were averaged and subjected to statistical analysis of all the characters. To test the variation among various genotypes of sweet potato, analysis of variance was carried [14]. Mean value of all the seven morphological characters for each plot of all the three replications was used for simple matching using UPGMA algorithm [15]. Principal component analysis (PCA) was used to depict nonhierarchical relationships among the genotypes. Eigen values and eigenvectors were calculated by the Eigen program using a correlation matrix as input (calculated using standardized morphological data), and a 2D plot was used to generate the two-dimensional PCA plot from NTSYS-pc 2.2 [16].

2.3. Biochemical analysis

2.3.1. *Total carotenoids (mg 100 g^{-1} fresh weight)*

Flesh was cut into small pieces longitudinally and mixed with 80% aqueous acetone for 2 hours at 50°C using an orbital shaker. Then it was filtered through Whatman paper. Filtrate was kept at −20°C prior to analysis. Five ml of sample extract was mixed with 5 ml distill water and 1 ml of mix (hexane/acetone/methanol) (50/25/25 v/v). Sample was kept at centrifuge at 3000 rpm for 10 min. The absorbance of upper layer was measured at 450 nm. Total carotenoids of the sample were calculated as μg 100 g^{-1} [17]:

$$\text{Total carotenoids } (\mu g\ 100\ g^{-1}) = \frac{A \times \text{Volume of the extract}}{A^{x} \times \text{Sample weight}}$$

where A = absorbance and A^x = absorbance coefficient (2505).

2.3.2. *Total antioxidant (mg 100 g^{-1} fresh weight)*

Five gram tuber was extracted with 20 ml of 60% methanol (0.1% HCl) and kept overnight. Then it was centrifuged at 10,000 rpm for 15 min at 10°C. The supernatant was taken for analysis. Hundred µl of methanolic extract was mixed with 3 ml of solution (1.2 M sulfuric acid, 46 mM sodium phosphate, and 8 mM ammonium molybdate) and was incubated for 90 min at 95°C in water bath. It was then allowed to cool down at room temperature. Reading of plant sample was read using spectrophotometer at a wavelength of 695 nm, and ascorbic acid was taken as standard. Standard curve was plotted with the absorbance readings of standard and plant sample, which gave value of total antioxidant in mg 100 g^{-1} [17].

2.3.3. *Total phenol (mg 100 g^{-1} fresh weight)*

Phenolic compounds are well-known phytochemicals found in all plants. They consist of simple phenols, benzoic and cinnamic acid, coumarins, tannins, lignins, lignans, and flavonoids. The total phenolics in sweet potato extracts were estimated by Folin-Ciocalteu colorimetric method. One gm of powdered sample was added in 80% ethanol followed by centrifugation at 10,000 rpm for 20 min, this step was repeated twice. Supernatant was taken and evaporated for dryness. The residue was dissolved in 5 ml of distilled water. Diluted sample extract (1 ml) was added 3.0 ml of 20-fold diluted Folin-Ciocalteu reagent and 2 ml of 20% (w/v) Na_2CO_3. The mixture was incubated in a water bath at 50°C for 1 min and allowed to cool. The absorbance was measured at 650 nm and used to calculate total phenolics content using a standard curve [17].

2.4. Molecular analysis

The genomic DNA was extracted from young leaves (3-week-old plantlets) using CTAB method [18] with slight modifications. Totally, 20 RAPD primers were used for PCR amplification out of which 15 primers produced polymorphic, consistent, and reproducible banding pattern. In brief, reproducible and clear banding patterns were obtained in a reaction mixture of 20 µl containing 1X reaction buffer, 1 unit of *Taq* DNA polymerase, 200 µM each of dNTPs mix, 20 pmol of primer, and 50 ng of template DNA. PCR amplification in the thermocycler (programmable thermal cycler from BIORAD™ International) was programmed for an initial denaturation step of 5 min at 94°C, followed by 40 cycles of denaturation (94°C, 1 min), annealing (37°C, 1 min), and extension (72°C, 1 min) followed by a final extension of 72°C for 5 min and a hold temperature of 4°C. The amplified products were electrophoresed on 1.5% agarose (Sigma chemicals Co. Ltd. India) gels in TAE buffer at 50 volts for 3 hours. The electrophoresed gels were visualized under UV transilluminator and photographed using gel documentation system [19]; 100 bp ladder and 1 Kb DNA ladder were used as standard (Bangalore, Genei, India).

2.5. Data analysis

The amplicons obtained from different RAPD markers were scored based on the presence (taken as 1) or absence (taken as 0) of bands for each primer. Accordingly, a rectangular binary matrix is obtained, and statistical analysis was performed using the NTSYS-pc version 2.02e [16]. The pair-wise association coefficient was calculated from qualitative data using Jaccard's similarity coefficient (by means of SIMQUAL procedure of NTSYS-pc), and the cluster analysis was performed (by means of SAHN procedure of NTSYSpc) via unweighted pair group method with arithmetic averages (UPGMA) to develop a dendrogram [15]. A two-dimensional and three-dimensional principal component analysis (PCA) was constructed to provide another means of testing the relationship among the cultivars using the Eigen program (NYSTS-pc). The effective number of alles, Nei's genetic diversity/expected number of heterozygosity, and Shannon's Information index were computed using Popgene software. A two-dimensional and three-dimensional principal component analysis (PCA) was constructed to provide another means of testing the relationship among the cultivars using the Eigen program NTSYS-pc version 2.02 [15].

2.6. Association between genetic, morphological, and biochemical diversity

The cophenetic correlation was calculated to find the degree of association between the original similarity matrix and the tree matrix in morphological, molecular, and biochemical analyses. Using the Mantel test [20], a comparison between all the methods was performed for the accessions for which both data sets were available by calculating the correlation between the three data sets in NTSYS-pc. Using the same software, PCA was also carried out to identify any genetic association among the genotypes.

3. Results and discussions

3.1. Morphological analysis

Morphological characterization is regarded as the first step in description and classification of any germplasm. A sound knowledge of various morphological traits in the breeding material helps classification, identification, naming, and documentation of the entries in a crop. These hasten the process of utilization of genetic material for crop improvement programs. The data of morphological characters were subjected to analysis of variance (ANOVA) for randomized block design (RBD). The mean square values due to genotypes were found significant for all the traits, thereby indicating substantial amount of variability among the genotypes (**Table 1**). The mean squares due to replication were found significant for all the characters, namely, storage root number per plant, individual root weight (g), storage root fresh root per plant, storage root dry yield per plant (g), storage root yield per plot (kg), storage root length (cm), which had high value of variance indicating that the diversity existed, which can contribute to improvement of the crop. The mean performances of different *Ipomoea batatas* L. genotypes for different characters are presented in **Table 1**.

The genotypes Gauri and 187017 showed maximum root number per plant, that is, 7 and C-71, SI-60, SV-71, and SREE VARDHINI, that is, 2 with general mean of 3.86. The perusal of mean table revealed that the test germplasm V-11 was found to be the lowest in root weight (35 g), whereas C-71(1100 g) was the highest among the test germplasm, the germplasm C-71 exhibited highest values of storage root fresh yield per plant (2200 g), while minimum in V-10 (160 g). There was significant difference in yield for all the germplasm with an average value of 433.71 g. The germplasm

SN	Genotype	Storage root number/plant	Individual root wt (g)	Storage root fresh yield/ plant (g)	Storage root dry yield/ plant (g)	Storage root yield/plot (kg)	Storage root length (cm)
1	Gauri	7.00	60.00	420.00	134.40	3.36	20.00
2	ST-10	5.00	65.00	325.00	113.75	2.84	22.00
3	V-10	4.00	40.00	160.00	56.00	1.40	16.00
4	SP-2	4.00	55.00	220.00	77.00	1.92	25.00
5	V-11	5.00	35.00	175.00	51.25	1.28	17.00
6	C-71	2.00	1100.00	2200.00	540.00	13.50	33.33
7	Kamla Sundari	4.00	120.00	480.00	129.60	3.24	10.00
8	Sree Nandini	3.00	150.00	450.00	157.50	3.93	25.00
9	SP-1	3.00	75.00	225.00	78.75	1.95	26.00
10	CIPSWA-2	3.00	80.00	240.00	84.00	2.10	27.00
11	SI-60	2.00	100.00	200.00	70.00	1.75	25.00
12	SV-71	2.00	110.00	220.00	77.00	1.92	23.00
13	440127	3.00	91.67	285.00	99.75	2.50	28.00
14	V-7	3.00	60.00	180.00	63.00	1.57	19.00
15	Khangudu	5.00	144.00	720.00	129.60	3.25	20.00
16	Pol-19-8-2	4.00	124.00	496.00	173.60	4.34	26.00
17	MPUAT-6	4.00	103.00	412.00	144.20	3.60	28.00
18	Varsha	4.00	50.00	200.00	54.00	1.35	8.00
19	ST-14	5.00	110.00	550.00	184.33	4.60	12.00
20	Sree Vardhini	2.00	125.00	250.00	85.00	2.12	17.00
21	187017	7.00	100.00	700.00	199.50	4.99	13.00
	GM	3.86	137.98	433.71	128.68	3.21	20.97
	SE	0.25	7.92	18.04	2.69	0.16	1.12
	CD5	0.71	22.64	51.55	7.67	0.46	3.20
	CD1	0.96	30.31	69.02	10.27	0.61	4.29
	CV	11.21	9.94	7.20	3.61	8.66	9.25

Table 1. Mean values of morphological characteristics observed for sweet potato.

V-11 had minimum (51.25 g) among the test germplasm, while C-71 had maximum (540 g) for dry yield per plant. Storage root yield per plot was maximum in the germplasm C-71 (13.50 kg), whereas V-11 had the minimum value of root yield/plot (1.28 kg). The highest storage root length (cm) was observed in germplasm C-71 (33.33 cm), while the minimum was VARSHA (8 cm).

SN	Characters	Replication	Genotype	Error
	d.f.	[2]	[20]	[40]
1	Storage root number/plant	0.5119	6.3857**	0.1869
2	Individual root weight (g)	318.7778	149072.5125**	188.2
3	Storage root fresh yield/plant (g)	3105.1904	577303.5500**	976
4	Storage root dry yield/plant (g)	53.9544	32674.7156**	21.63
5	Storage root yield/plot (kg)	0.0237	20.4307**	0.07747
6	Storage root length (cm)	4.0635	133.0635**	3.763

*, **, significant at 5% and 1% level of significance, respectively.

Table 2. Analysis of variance (ANOVA) for morphological traits of experimental data.

The analysis of variance was done for six characters studied, and their mean square values are presented in **Table 2**. The mean squares and mean squares due to replication were also found to be highly significant for all the characters. The overall analysis of ANOVA indicated the presence of high genetic variability in the experimental material, which can be further exploited for sweet potato improvement. Elameen et al. (2011) studied the phenotypic diversity of morphological plant and root descriptor traits in 105 sweet potato germplasms using 27 phenotypic characters using ANOVA. The analysis of variance (ANOVA) revealed highly significant variation among the accessions for 21 out of the 27 characters studied [21].

Comparative analysis of seven morphological characters revealed low level of variation. Pairwise similarity among the genotypes of *Ipomoea batatas* L. ranged from 0.01 to 0.50 with an average of 0.176 based on morphometric data. A dendrogram generated from morphometric data grouped all 21 genotypes into three major clusters (**Figure 1**). The first cluster involved seven germplasms, namely, Gauri, V-7, Sree Nandini, SP-1, CIPSWA-2, 440127, and 187017-1 at similarity coefficient of 0.01. This cluster was further divided into two subclusters A and B. Subcluster A included six germplasms: Gauri, V-7, Sree Nandini, SP-1, CIPSWA-2, and 440127 at similarity coefficient of 0.08. Subcluster B comprises only one germplasm, viz., 187017-1 at similarity level of 0.03, which was most distinct from all other germplasms. The second cluster comprised four germplasms, namely, ST-10, V-11, Khangadu, and ST-14 at similarity value of 0.161. The third cluster was the biggest that included 10 germplasms, namely V-10, Kamla Sundari, POL-19-8-2, MPUAT-6, Varsha, SP-2, SV-71, SI-60, C-71, and Sree Vardhini at similarity value of 0.04. This cluster was further divided into two subclusters C and D. Subcluster C included five germplasms: V-10, Kamla Sundari, POL-19-8-2, MPUAT-6, and Varsha at similarity coefficient of 0.161. Subcluster D was further divided into D1 and D2. Subcluster D1 included SP-2, SV-71, and SI-60 at similarity coefficient of 0.11. SP-2 and SV-71 found to be morphologically quite similar at similarity value of 0.50. Subcluster D2 comprised germplasms C-71 and Sree Vardhini.

Yada et al. (2010) assessed morphological characterization of 1256 Ugandan sweet potato germplasms and grouped the 1256 accessions into 20 major clusters, with the number of

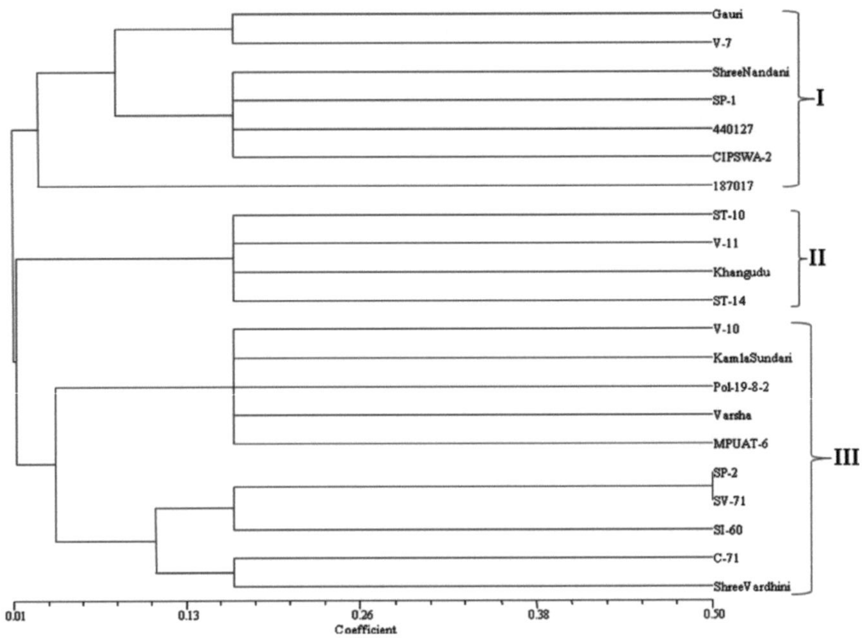

Figure 1. Dendrogram generated for 21 *Ipomoea batatas* L. germplasm using UPGMA cluster analysis based on morphological characters.

accessions per cluster ranging from 15 to 166 [22]. Similarly, Norman et al. (2014) also evaluated the diversity within sweet potato germplasm using factor and cluster analyses and revealed eight clusters at distance coefficient of 0.80 [23].

The two-dimensional plot generated from PCA showed three groups that were found to be similar to the clustering pattern of the UPGMA dendrogram. In the 2D plot, genotype 187017 was found distinct as like UPGMA dendrogram. Genotype C-71 was found along with Sree Vardhini in UPGMA dendrogram, whereas it was most distinct in 2D plot and made separate group (**Figure 2**). The analysis gave five principal components (PCs), out of which the first four principal components contributed 99.81% of the total variability.

The first three principal components accounted for 99.12% of the total variability, in which the highest variation was contributed by the first component (68.48%), followed by second (22.07%) and third components (8.56%). The first PC was influenced by the characteristics of the storage root yield/plot (kg), storage root dry yield/plant (g), storage root number/plant, storage root fresh yield/plant (g), and storage root length (cm). In the second PC, the traits contributing to the total variability were storage root length (cm), storage root number/plant, individual root weight (g), storage root yield/plot (kg), and storage root dry yield/plant (g). Third PC was influenced by the characteristics of the storage root number/plant, storage root fresh yield/plant (g), and storage root length (cm). In the fourth PC, the traits contributing to the total variability were the storage root number/plant, individual root wt (g), storage root yield/plot (kg), and storage root dry yield/plant (g) (**Table 3**).

The results presented in the present investigation are in support with the earlier studies. Moulin et al. (2012) characterized of 46 sweet potato landraces using morphological descriptors

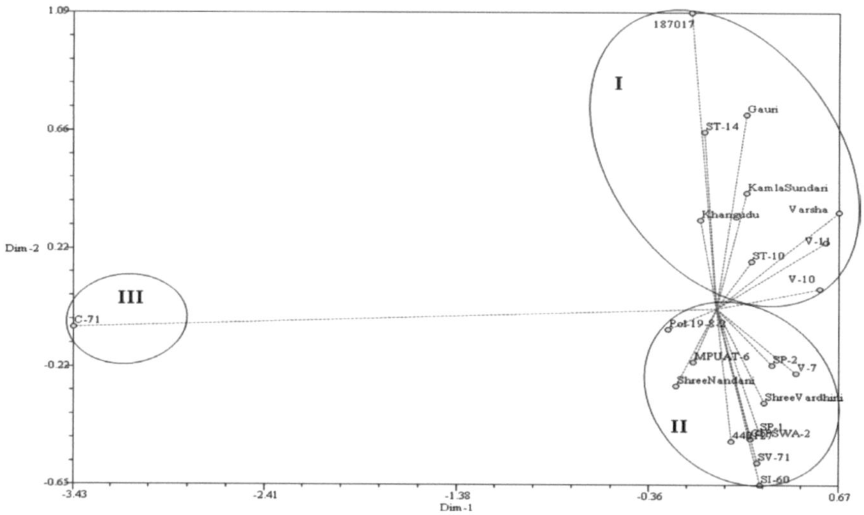

Figure 2. 2D plot generated for 21 *Ipomoea batatas* L. genotypes using UPGMA cluster analysis based on morphological characters.

	PC-1	PC-2	PC-3	PC-4
Storage root number/plant	0.611	0.1762	0.1481	0.3114
Individual root wt (g)	−0.3443	0.0796	−0.0025	0.009
Storage root fresh yield/plant (g)	0.0484	−0.0704	0.0235	−0.0048
Storage root dry yield/plant (g)	0.1317	0.0496	−0.0955	0.0034
Storage root yield/plot (kg)	0.1316	0.0487	−0.0949	0.0032
Storage root length (cm)	0.323	0.5878	0.103	−0.0388

Table 3. Detail of principal components based on morphological traits.

and reported that the morphological characterization was efficient to detect genetic variability among accessions [24]. Amoatey et al. (2016) also reported the significant genetic variability among the 20 accessions of sweet potato studied based on the agromorphological characters, and the hierarchical cluster analysis grouped these accessions into two clusters at a genetic similarity index of 61.60% [25].

3.2. Biochemical analysis of tubers

3.2.1. *Total carotenoid (mg 100 g^{-1} fresh weight)*

The data presented in **Table 4** indicated that germplasm studied has significant differences for total carotenoids. The range for total carotenoids was 0.76–9.24 mg 100 g^{-1} fresh weight. The mean for total carotenoids was 4.84 mg 100 g^{-1} fresh weight. The maximum total carotenoids were recorded in tubers of "ST-14" (9.24 mg 100 g^{-1} fresh weight), while minimum was recorded in "SREE NANDNINI" (0.76 mg 100 g^{-1} fresh weight).

S.No.	Germplasm	Total carotenoids mg 100 g^{-1} fresh weight	Total Antioxidant mg 100 g^{-1} fresh weight	Total Phenols mg 100 g^{-1} fresh weight
1	Gauri	6.6	0.21	1.28
2	ST-10	1.01	0.28	1.07
3	V-10	1.02	0.18	1.32
4	SP-2	5.4	0.21	1.03
5	V-11	1.12	0.23	1.18
6	C-71	7.91	0.23	1.19
7	Kamla Sundari	6.46	0.2	1.25
8	Sree Nandini	0.76	0.26	1.38
9	SP-1	6.15	0.16	1.18
10	CIPSWA-2	6.91	0.23	0.94
11	SI-60	7.16	0.19	1.36
12	SV-71	1.17	0.19	1.3
13	440127	5.45	0.23	1.34
14	V-7	2.6	0.2	1.35
15	Khangudu	4.55	0.21	0.98
16	Pol-19-8-2	7.95	0.23	1.37
17	MPUAT-6	1.1	0.23	1.21
18	Varsha	4.45	0.17	1.08
19	ST-14	9.24	0.25	1.05
20	Sree Vardhini	0.87	0.19	1.16
21	187017	6.3	0.19	1.03

Table 4. Total carotenoids, total antioxidants, and total phenols in sweet potato tubers.

Liu et al. (2008) reported that the orange-fleshed sweet potato had higher total carotenoids content than yellow-fleshed [26]. Eluagu et al. (2010) reported that utilizing orange-fleshed sweet potato in their raw (unbalanced) form retains nutrient more than in their processed form and total carotenoids in the 10 orange-fleshed clones ranged between 10.32 and 13.99 mg 100 g^{-1} of fresh weight [27].

3.2.2. *Total antioxidant (mg 100 g^{-1} fresh weight)*

The data presented in **Table 4** indicated that germplasm showed significant differences for total antioxidants. The mean value for total antioxidants was 0.08 mg 100 g^{-1} fresh weight, and it ranged from 0.16 to 0.28 mg 100 g^{-1} fresh weight. The maximum total antioxidants were observed in "ST-10" (0.28 mg 100 g^{-1} fresh weight) followed by "Sree Nandini"

(0.26 mg 100 g^{-1} fresh weight) and "ST-14" (0.25 mg 100 g^{-1} fresh weight). The minimum total antioxidants were observed in "SP-1" (0.16 mg 100 g^{-1} fresh weight). Padda and Picha (2007) quantified the antioxidant activity and phenolic content of sweet potato roots and leaves of different sizes and ages [28]. Khurnpoon and Rungnoi (2012) also estimated the total phenol content and antioxidant activities of 36 sweet potato (*Ipomoea batatas*) cultivars with distinctive flesh color (white, yellow, orange, and purple) grown in Thailand [29].

3.2.3. Total phenol (mg 100 g^{-1} fresh weight)

The data presented in **Table 4** indicated that germplasm showed significant differences for total phenols. The mean value for total phenols was 1.19 mg 100 g^{-1} fresh weight, and it ranged from 0.94 to 1.38 mg 100 g^{-1} fresh weight. The maximum total phenols were observed in "SREE NANDNI" (1.38 mg 100 g^{-1} fresh weight) followed by "POL-19-8-2" (1.37 mg 100 g^{-1} fresh weight) and "SI-60" (1.36 mg 100 g^{-1} fresh weight). The minimum phenols were observed in "CIPSWA-2" (0.94 mg 100 g^{-1} fresh weight). Vyas et al. (2014) evaluated total phenolic content (TPC) of four successive extracts of various parts of *Nyctanthes arbortristis* Linn. TPC revealed that all extracts act as radical scavengers possibly due to presence of polyphenolic compounds and concluded that *Nyctanthes arbortristis* Linn. exhibited strong antioxidant activity and could serve as potential therapeutic plant for various diseases [30].

3.3. Molecular analysis

Out of 15, 10 RAPD primers showed variable degree of amplification and generated total 96 bands, which were polymorphic (**Figure 3**; **Table 5**). Only those fragments that were consistently amplified were considered for analysis. **Table 5** combines the comparative information about total number of fragments with base pair obtained by all the primers in all sweet potato genotypes. The advent of the RAPD provided a competent method to detect DNA polymorphism and generate a large number of molecular markers for genomic applications [31].

RAPD markers are easy and rapid and have the advantage of no prior knowledge of genome sequence. RAPD technique can be used in laboratories with limited resources but requires optimization for reproducible results for each species under research. Once the reaction conditions have been optimized, the technique is consistent and instructive. Silva et al. (2014) studied genetic diversity using RAPD markers. Marker showed that the collection had a high level of polymorphism. By UPGMA, they separated three groups of genotypes and identified two reconstructed populations by structure software [32].

A dendrogram was constructed using similarity matrix values determined from RAPD data for 21 sweet potato genotypes using UPGMA of NTSYS-pc software (Rholf, 2000). The similarity coefficient for different genotypes was in the range of 0.65–0.83. The average similarity across all the genotypes was found to be 0.74. A dendrogram generated from molecular data grouped all 21 genotypes into five major clusters (**Figure 4**).

The first cluster includes four genotypes, namely, Gauri, CIPSWA-2, Sree Vardhini, and Sree Nandini in which Sree Nandini was most distinct with all the remaining genotypes with similarity value of 0.66. Second cluster comprises only two genotypes like Kamla Sundari and

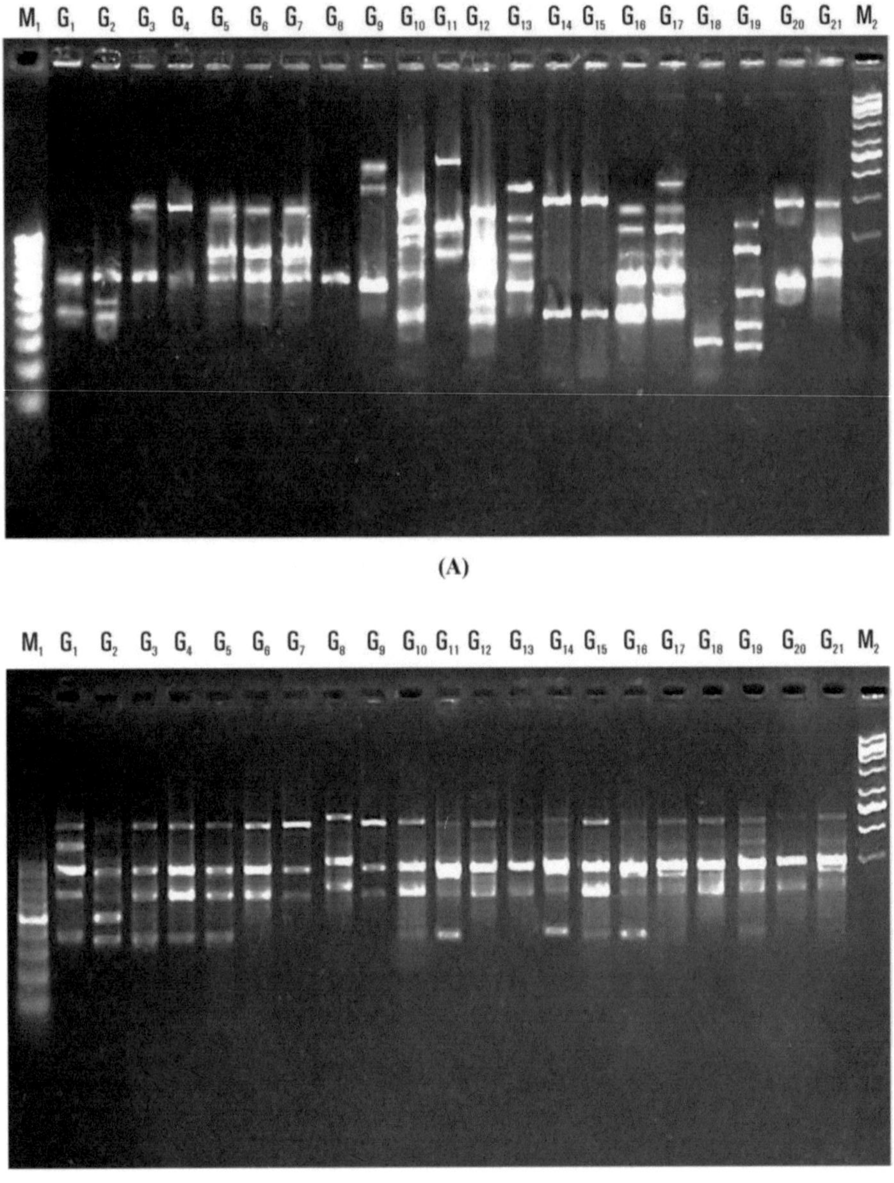

Figure 3. RAPD profiles of *Ipomoea batatas* L. DNA samples obtained from different germplasms using primer (A) OPE-04 and (B) OPE-03.

187017 with similarity value of 0.74. Third cluster comprises two genotypes, namely, SV-71 and Pol-19-8-2 with a similarity value of 0.73. The fourth cluster includes six genotypes, viz., V-11, SP-2, MPUAT-6, SP-1, ST-10, and V-7. The V-11 genotype was the most diverse from other genotypes in this group with similarity value of 0.73. The fifth cluster was the biggest one and comprised eight genotypes, namely, C-71, 440127, Varsha, Mahangudu, V-10, SI-60, and ST-14. Within this cluster, C-71 and 440127 genotypes were the most similar to each other

S.No	Primers code	Total number of bands (a)	Total number of polymorphic bands (b)	Polymorphism % (b/a*100)
1	OPE-03	10	10	100
2	OPE-04	12	12	100
3	OPA-07	10	10	100
4	OPA-11	8	8	100
5	OPM-03	10	10	100
6	OPM-06	11	11	100
7	OPA-05	7	7	100
8	OPA-09	11	11	100
9	OPP-10	10	10	100
10	OPD-05	7	7	100
Total		**96**	**96**	**100**

Table 5. Polymorphism information of RAPD primers analyzed.

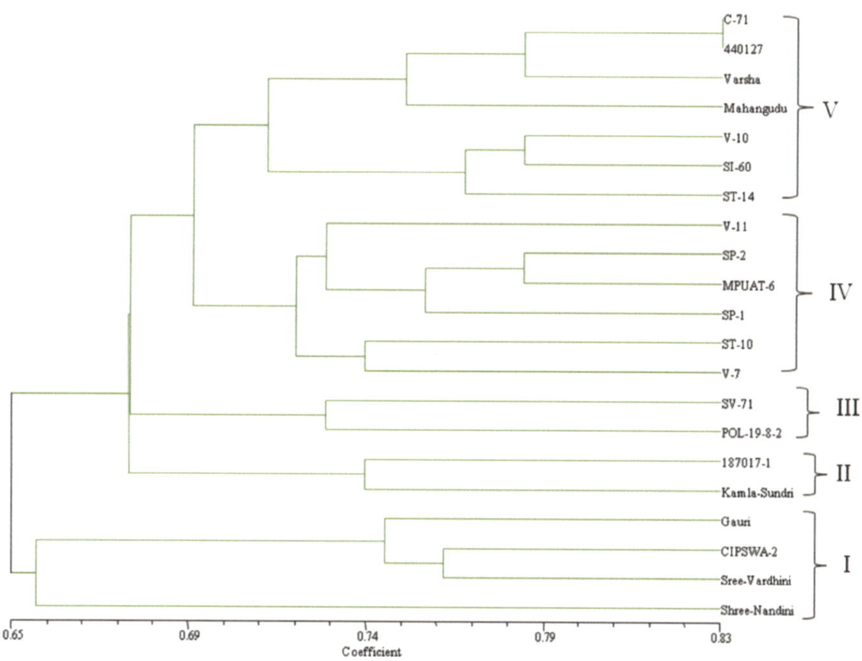

Figure 4. Dendrogram generated for 21 *Ipomoea batatas* L. germplasm using UPGMA cluster analysis based on RAPD marker.

with similarity value of 0.83 (**Figure 3**). Wang et al. (2009) assessed genetic distance and cluster analysis in 30 sweet potatoes based on the 26 RAPD primers and showed that the genetic distance among the 30 sweet potato varieties ranged from 0.0390 to 0.4306 with an average of

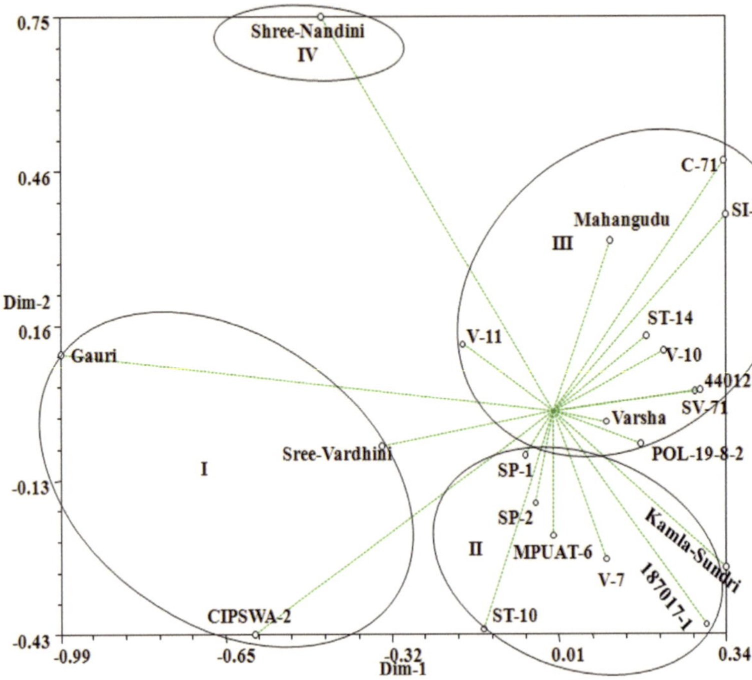

Figure 5. 2D plot generated for 21 *Ipomoea batatas* L. germplasms using UPGMA cluster analysis based on RAPD marker.

0.3086 [33]. The dendrogram based on RAPD markers indicated that the sweet potato varieties coming from the same regions or having the same parents were clustered into the same groups.

The two-dimensional plot generated from PCA showed four groups that were found to be similar to the clustering pattern of the UPGMA dendrogram (**Figure 5**). In the 2D plot, genotype Sree Nandini was found most distinct, whereas it was present along with Sree Vardhini in UPGMA dendrogram.

The analysis gave 19 principal components (PCs), out of which the first 10 principal components contributed 72.42% of the total variability. The first three principal components accounted for 30.14% of the total variability, in which the highest variation was contributed by the first component (12.30%), followed by the second (9.52%) and third components (8.31%). Similar clustering pattern was detected by Moulin et al. (2012) using eight RAPD primers. That results revealed that these eight primers with 44 sweet potato accessions generated a total of 93 scorable fragments, 88 of which (94.6%) are polymorphic [34]. Genetic relationship among sweet potato genotypes was also visualized by performing PCA based on RAPD data. The results of PCA were comparable to the cluster analysis with minor differences. Genotypes grouped within the same cluster in the dendrogram were also occupying the same position in two-dimensional and three-dimensional scaling based on molecular data.

3.4. Association between molecular and morphological diversity

Cluster analysis was performed for both morphological and molecular data using the unweighted pair group method using arithmetic mean (UPGMA) algorithm, from which dendrograms depicting the similarity among germplasm were drawn and plotted using NTSYS-pc. The correlation was calculated to find the degree of association between the original similarity matrix in both morphological and molecular analyses. Using the Mantel test, a comparison between both methods was performed for the genotypes and showed moderate level of correlation between molecular and morphological data, i.e., r = 0.21279. This finding is agreed with that of Elameen et al. (2011) who studied the phenotypic diversity of morphological plant and root descriptor traits in 105 sweet potato germplasms using 27 phenotypic characters. Cluster analysis was conducted using the unweighted pair group method with arithmetic mean (UPGMA) [21].

3.5. Cumulative data analysis of morphology, biochemical, and molecular markers

Pairwise similarity among the genotypes ranged from 0.58 to 0.77 with an average of 0.67 based on combined morphometric, biochemical, and molecular markers data. The highest similarity (77%) was observed between C-71 and 440127 genotypes, whereas the lowest was observed between Sree Nandini and 187017-1 with a similarity value of 0.50. A dendrogram

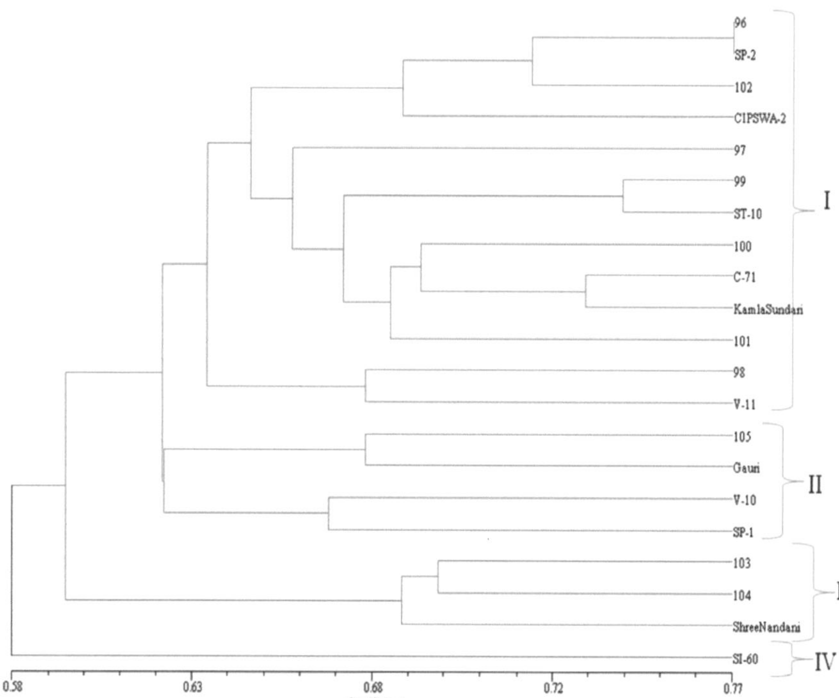

Figure 6. Dendrogram generated for 21 *Ipomoea batatas* L. germplasms using UPGMA cluster analysis based on cumulative morphological, biochemical, and molecular data.

based on combined morphometric, biochemical, and molecular marker data clustered all 21 genotypes into four major clusters (**Figure 6**).

The first cluster was the biggest one and comprised 13 genotypes, namely, C-71, 440127, Varsha, Mahangudu, V-11, V10, SI-60, SP-2, MPUAT-6, ST-14, SP-1, ST-10, and V-7. Within this cluster, C-71 and 440127 were the most similar morphologically, biochemically, and genetically, showing a similarity value of 0.77. In this group, V-11 was distinct from the other genotypes, with a similarity value of 0.65. The second cluster comprised four genotypes: 187017-1, Kamla Sundari, SV-71, and Pol-19-8-2. Third cluster comprised three genotypes, namely, Gauri, CIPSWA-2, and Sree Vardhini. Within this cluster, Gauri and CIPSWA-2 were most similar to each other with a similarity value of 0.70. Fourth cluster was smallest one and comprised only one genotype, namely, Sree Nandini, which was highly distinct from other genotypes with a similarity coefficient of 0.58. Based on Mantel Z-statistics (Mantel, 1967), the correlation coefficient (r) was estimated as 0.14. This value was considered a good fit of the UPGMA cluster pattern to the cumulative morphological, biochemical, and molecular data.

The 2D plot generated from the PCA of the combined morphological, biochemical, and molecular data (**Figure 7**) also supported the clustering pattern of the UPGMA dendrogram. In the 2D plot, genotype SP-1 was grouped in cluster III, whereas in UPGMA clustering, it was grouped in cluster I. However, in 2D plot, Sree Nandini genotype grouped with Gauri, CIPSWA-2, and Sree Vardhini genotypes, whereas it was in separate clusters in UPGMA pattern.

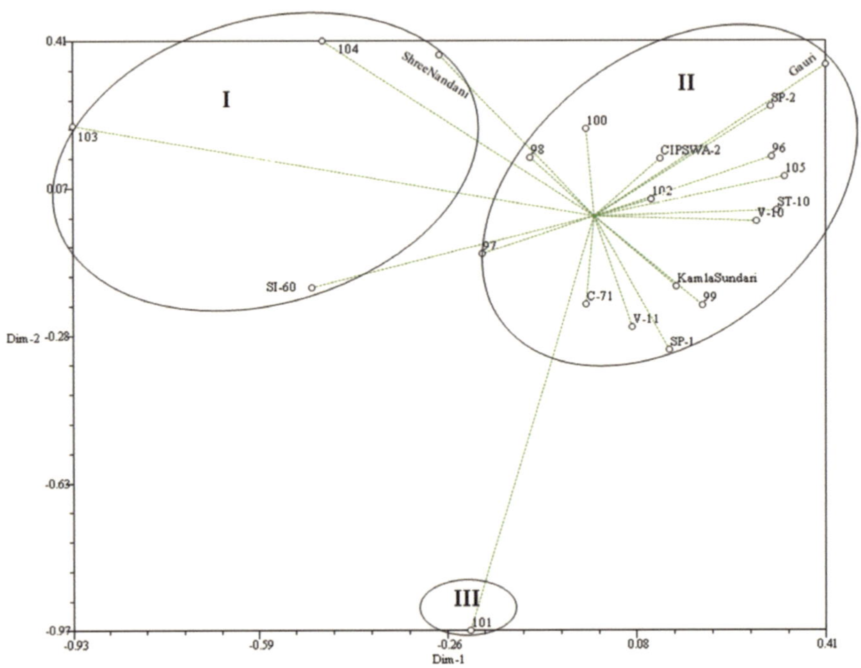

Figure 7. 2D plot generated for 21 *Ipomoea batatas* L. germplasms using UPGMA cluster analysis based on cumulative morphological, biochemical, and molecular data.

The analysis gave 19 PCs, out of which the first 10 PCs contributed 72.34% of the total variability of the analyzed genotypes. The first five PCs accounted for 44.88% of the total variability; the first three accounted for 30.01% of the variance, in which maximum variability was contributed by the first component (11.40%), followed by the second (9.63%) and third (8.96%) components. Kaur et al. (2016) reported the UPGMA dendrogram based on the combined morphological and molecular markers in which the 23 mungbean genotypes were divided into three main clusters, showing a close genetic relationship, which might be due to their close genetic bases [35].

4. Conclusion

Findings of the present study revealed that 21 sweet potato germplasms were moderate to high diversity based on molecular, biochemical, and morphological assessment approaches. The results obtained will serve as a guide for the basis of genotype management and crop improvement programs. Designing effective breeding programs is largely dependent on understanding the genetic diversity of the relevant germplasms. Here, we reported our detailed analysis of representative sweet potato accessions cultivated using morphological, biochemical, and molecular markers. Our results demonstrated significant genetic diversity in the orange flashed sweet potato germplasm collection. Although sweet potato is highly heterozygous, the limited scope of parent selection in breeding also affected the genetic diversity of advanced varieties of sweet potato. To create new hybrid varieties with new alleles and increased genetic diversity, sweet potato accessions with a wide genetic background that includes introduced varieties should be used in breeding programs.

From the discovery of first molecular marker, there was a continuous development in the molecular markers technology from RFLP to SNPs and a diversity of array-technology-based markers. Advancements in the nanopore-based sequencing technologies have led to development of low-cost sequencing with high throughput. In spite of the presence of these highly advanced molecular marker tools and techniques, the outcomes of such technologies are yet to come due to inaccurate phenotyping. Application of molecular marker technologies also lies in the areas of plant biology like systematics, population genetics, evolutionary biology and conservation genetics, genomics, identification of the wild progenitors of domestic species, and the establishment of geographic patterns of genetic diversity. The success of molecular marker technology for bringing crop improvement depends on the positive interaction between plant breeders and biotechnologists.

Acknowledgements

Financial assistance from AINP on "Soil Biodiversity and Biofertilizers" and RKVY project on "Validation of important crop varieties through DNA fingerprinting" is highly acknowledged.

Conflict of interest

All the authors declare that there is no conflict of interest.

Author details

Anubhuti Sharma[1], Devendra Jain[1*], Sunil Kumar Khandelwal[1], Ravish Chaudhary[2], Kapil Dev Ameta[3] and Abhijeet Singh[4]

*Address all correspondence to: devendrajain@mpuat.ac.in

1 Department of Molecular Biology and Biotechnology, Rajasthan College of Agriculture, Maharana Pratap University of Agriculture and Technology, Udaipur, India

2 Division of Seed Science and Technology, ICAR-Indian Agricultural Research Institute, New Delhi, India

3 Department of Horticulture, Rajasthan College of Agriculture, Maharana Pratap University of Agriculture and Technology, Udaipur, India

4 Department of Biosciences, Manipal University, Jaipur, India

References

[1] Wadl PA, Olukolu BA, Branham SE, Jarret RL, Yencho GC, Jackson DM. Genetic diversity and population structure of the USDA sweetpotato (*Ipomoea batatas*) germplasm collections using GBSpoly. Frontiers in Plant Science. 2018;**21**:1166. DOI: 10.3389/fpls.2018.01166. eCollection 2018

[2] Horticultural Statistics at a Glance. Horticulture Statistics Division, Department of Agriculture, Ministry of Agriculture & Farmers Welfare. New Delhi, India: Government of India; 2017. p. 481

[3] Palmer JK. Carbohydrates in sweetpotato. In: Villareal R, Griggs T, editors. Proceedings of the First International Symposium; 23-27 March 1981; Taiwan. China: AVRDC; 1982. pp.137-138

[4] Wang A, Li R, Ren L, Gao X, Zhang Y, Ma Z, et al. A comparative metabolomics study of flavonoids in sweet potato with different flesh colors (*Ipomoea batatas* (L.) lam). Food Chemistry. 2018;**260**:124-134

[5] Ghosh SP, Ramanujan T, Jos JS, Moorthy SN, Nair RG. Tuber Crops. New Delhi: Oxford IBH Publishing; 1988. 146 p

[6] Byers T, Perry G. Dietary carotenes, vitamin C and vitamin E as protective antioxidants in human cancers. Annual Review of Nutrition. 1992;**12**:139-159

[7] Montilla EC, Hillebrands S, Winterhalter P. Anthocyanins in purple sweet potato (*Ipomoea batatas*) varieties. Fruits, Vegetable and Cereal Science and Biotechnology. 2011; **5**(2):19-24

[8] Woolfe JA, editor. Sweet Potato an Untapped Food Resource. UK: Cambridge University Press; 1992. 643 p

[9] Gester H. Anti carcinogenic effect of common carotenoid. International Journal for Vitamin and Nutrition Research. 1993;**63**:93-121

[10] Powell W, Morgante M, Andre C. The comparison of RFLP, RAPD, AFLP and SSR (microsatellite) markers for germplasm analysis. Molecular Breeding. 1996;**2**:225-238

[11] Russel JR. Direct comparison of levels of genetic variations among barley accessions detected by RFLPS, SSRS and RAPDS. Theoretical and Appliied Genetics. 1997;**95**:714-722

[12] Fahima T, Sun GL, Beharav A, Krugman T, Beiles A, Nevo E. RAPD polymorphism of wild emmer wheat populations, *Triticum dicoccoides* in Israel. Theoretical Applied Genetics. 1999;**98**:43-47

[13] Subudhi PK, Huang N. RAPD mapping in a doubled haploid population of rice (*Oryza sativa* L.). Hereditas. 1999;**130**:2-9

[14] Panse VG, Sukhatme PV. Statistical Methods for Agricultural Workers. New Delhi: Indian Council of Agricultural Research; 1975. 361 p

[15] Sneath PHA, Sokal RR. Numerical Taxonomy. San Francisco: Freeman; 1973. 575 p

[16] Rohlf FJ. NTSYS-PC Numerical Taxonomy and Multivariate Analysis System, Version 2.02. Setauket, New York: Exeter Software; 2000

[17] Thimmiah SR. Estimation of chlorophyll. In: Standard Method of Chemical Analysis. New Delhi: Kalyani Publisher; 1999. 307 p

[18] Murray M, Thompson WF. Rapid isolation of high molecular weight plant DNA. Nucleic Acids Research. 1980;**8**:4321-4326

[19] Sambrook J, Fritschi EF, Maniatis T. Molecular Cloning: A Laboratory Manual. 2nd ed. New York, USA: Cold Spring Harbor Laboratory Press; 1989. 1546 p

[20] Mantel NA. The detection of disease clustering and a generalized regression approach. Cancer Research. 1967;**27**:209-220

[21] Elameen A, Larson A, Klemsadel SS, Fjellheim S, Sundheim L, Mosalla S, et al. Phenotypic diversity of plant morphological and root descriptor traits within a sweet potato, *Ipomoea batatas* (L.) germplasm collection from Tanzania. Genetic Resources and Crop Evolution. 2011;**58**:397-407

[22] Yada B, Tukamuhabwa P, Wanjala B, Kim DJ, Skilton RA, Alajo A, et al. Characterization of Ugandan sweetpotato germplasm using fluorescent labeled simple sequence repeat markers. Hortscience. 2010;**45**(2):225-230

[23] Norman PE, Beah AA, Samba JA, Tucker MJ, Benya MT, Fomba SN. Agro-phenotypic characterization of sweet potato (*Ipomoea batatas* L.) genotypes using factor and cluster analyses. Agricultural Science Research Journal. 2014;**4**(2):30-38

[24] Moulin MM, Rodrigues R, Gonçalvez LSA, Sundré CP, Pereira MGA. Comparison of RAPD and ISSR markers reveals genetic diversity among sweet potato landraces (*Ipomoea batatas* (L.) Lam.). Acta Scientiarum Agronomy. 2012;**34**:139-147

[25] Amoatey HM, Sossah FL, Ahiakpa JK, Quartey EK, Appiah AS, Segbefia MM. Phenotypic profiles of different accessions of sweet potato (*Ipomoea batatas* L. Lam) in the coastal savanna agro-ecological zone of Ghana. African Journal of Agricultural Research. 2016; **11**(26):2316-2328

[26] Liu SC, Lin JT, Yang DT. Determination of Cis-and Trans- α and β carotenoids in Taiwanese sweet potatoes (*Ipomoea batatas* L.) harvested at various times. Journal of Food Chemistry. 2008;**116**(3):605-610

[27] Eluagu E, Onimawo N, Ignitiatus A. Effect of processing on the antinutritional factors of orange fleshed sweet potato (*Ipomoea batatas* (L.) Lam) flours. Electronic Journal of Environmental, Agricultural and Food Chemistry. 2010;**9**(6):1000-1005

[28] Padda MS, Picha DH. Antioxidant activity and phenolic composition in 'beauregard' sweet-potato are affected by root size and leaf age. Journal of American Society Horticulture Science. 2007;**132**:447-451

[29] Khurnpoon L, Rungnoi O. The correlation between total phenol and antioxidant capacity of sweet potato (*Ipomoea batatas* L.) with varying flesh color. Acta Horticulturae. 2012;(945):413-419. DOI: 10.17660/ActaHortic.2012.945.56

[30] Vyas S, Kachhwaha S, Kothari SL. Comparituve analysis of the in vitro antioxidant activity and polyphenolic content of successive extracts of *Nyctanthes arbourtristis* Linn. International Journal of Pharmacy and Pharmaceutical Sciences. 2014;**6**:0975-1491

[31] Williams JGK, Kubelik AR, Livak KJ, Rafalski JA, Tingey SV. DNA polymorphisms amplified by arbitrary primers are useful as genetic markers. Nucleic Acid Research. 1990;**18**:6531-6535

[32] Silva AVCD, Andrade LNTA, Rabbani ARC, Nunes MUC, Pinheiro LR. Genetic diversity of sweet potatoes collection from Northeastern Brazil. African Journal of Biotechnology. 2014;**13**(10):1109-1116

[33] Wang Q, Zhang L, Wang B, Yin Z, Feng C, Wang Q. Sweet potato viruses in China. Crop Protection. 2009;**29**:110-114

[34] Moulin MM, Rodrigues R, Gonçalves LSA, Sudré CP, Santos MH, Silva JRP. Collection and morphological characterization of sweet potato landraces in north of Rio de Janeiro state. Horticultura Brasileira. 2012;**30**:286-292

[35] Kaur G, Joshi A, Jain D, Choudhary R, Vyas D. Diversity analysis of green gram (*Vigna radiata* (L.) Wilczek) through morphological and molecular markers. Turkish Journal of Agriculture and Forestry. 2016;**40**(2):229-240

Chapter 3

Bean Genome Diversity Reveals the Genomic Consequences of Speciation, Adaptation, and Domestication

Andrés J. Cortés, Paola Hurtado,
Mathew W. Blair and María I. Chacón-Sánchez

Additional information is available at the end of the chapter

http://dx.doi.org/10.5772/intechopen.80512

Abstract

Here we review whether genomic islands of speciation are repeatedly more prone to harbor within-species differentiation due to genomic features, such as suppressed recombination, smaller effective population size, and increased drift, across repeated hierarchically nested levels of divergence. Our discussion focuses on two species of *Phaseolus* beans with strong genepool and population substructure and multiple independent domestications each. We overview regions of species-associated divergence, as well as divergence recovered in within-species between-genepool comparisons and in within-genepool wild-cultivated comparisons. We discuss whether regions with overall high relative differentiation coincide with sections of low SNP density and with between-species pericentric inversions, since these convergences would suggest that shared variants are being recurrently fixed at replicated comparisons, and in a similar manner across different hierarchically nested levels of divergence, likely as the result of genomic features that make certain regions more prone to accumulate islands of speciation as well as within-species divergence. We conclude that neighboring signatures of speciation, adaptation, and domestication in *Phaseolus* beans seem to be influenced by ubiquitous genomic constrains, which may continue shaping, fortuitously, genomic differentiation at various other scales of divergence. This pattern also suggests that genomic regions important for adaptation may frequently be sheltered from recombination.

Keywords: genomic islands of speciation, genomic signatures of selection, adaptation, domestication syndrome, convergent evolution, gene flow, genomics constrains, GBS-derived SNP markers

1. Introduction: A strategy to discern among confounding causes of genomic divergence

Genomic signatures associated with species, genepools, and ecotypes' divergence can result from causes other than reduced gene flow, for example, random genetic drift and selection [1]. Moreover, the origin of the outlier variants from novel or standing genetic variation leads to distinctively different patterns of genomic divergence [2–4]. One approach that can help to distinguish these underlying causes of divergence is carrying out a replicated sampling of contrasting populations [5, 6]. If genetic drift rather than selection is responsible for the divergence, it is unlikely that signals of differentiation reappear consistently across replicates [5]. On the other hand, if selection acted on the same genetic variants at the replicated contrasting pairs, genomic regions with comparatively high divergence between individuals from contrasting populations should be identical at each of the replicated populations. Parallel selection on shared genetic variation should therefore lead to low divergence within populations and across replicates, in the exact genomic regions where equivalent variants are selected at each contrasting population [6]. Discerning among gene flow, genetic drift and selection as the cause of parallel genomic divergence are possible as long as there is some degree of replication considered in the sampling of contrasting populations.

The genomic landscape of divergence can also be influenced by differences in ancestral variation and recombination in the genome [7, 8]. Lineage sorting may be enhanced relative to background levels by a reduction in the effective population size (N_e) due to processes other than gene flow, like low recombination [8–10]. Since differentiation is further speeded up in low-recombining regions because of linked selection [11–13], the imprint caused by genomic features on the differentiation landscape should be ubiquitous across different levels of divergence. Therefore, besides a replicated sampling of contrasting populations, a hierarchical nested sampling across various scales of divergence is advisable in order to examine whether genomic islands of divergence may display differentiation due to suppressed recombination, smaller effective population size, and increased drift.

In order to discern among confounding causes of genomic divergence in a system with strong population structure and subjected to domestication, we suggest conducting the following analyses by taking advantage of a replicated hierarchical nested sampling across various scales of divergence:

A. Analyze whether F_{ST} outliers between species coincide with high F_{ST} values at within-species comparisons. This pattern is expected if genomic islands of speciation are repeatedly more prone to harbor within-species divergence as a result of limited recombination [8].

B. Assess whether the within-species between-genepool divergence F_{ST} profiles are similar among four available comparisons. This trend is expected if the same variants were selected as the result of similar selective pressures at multiple domestication events, but not if divergence outliers were due to population divergence, that is, genetic drift [5].

C. Assess whether the within-genepool wild-cultivated divergence F_{ST} profiles are similar among the available comparisons. This coincidence is expected if the same variants are selected as the result of parallel domestication but not if divergence is due to genetic drift [5].

Finally, we suggest exploring if regions of high F_{ST} co-localized with regions of low F_{ST} in within-population comparisons. Δ_{Div} can be used to analyze the difference between these two F_{ST} values in each window. Peaks in the Δ_{Div} statistic point to genomic regions that diverged as a result of parallel divergence from shared variation rather than due to novel variation evolving at each site [6].

2. Beans as a model system to study divergence across various scales of divergence in a replicated hierarchical nested framework

Phaseolus beans, with their striking genepool structure and multiple domestications, constitute an excellent model system [14, 15] to the approach described in the previous section and to explore to what extent genomic features, besides reduced gene flow and divergent selection, may lead to genomic divergence between (i.e., speciation islands) and within (i.e., during the natural colonization of new habitats as well as part of the domestication syndromes) species [16]. Common and lima beans are the only bean species with multiple domestications among the five domesticated species of *Phaseolus* [14]. Wild common bean (*P. vulgaris* L.) diverged from its sister species in the tropical Andes [17] and colonized South and Central America from its original distribution in Central America, originating what nowadays is known as the Andean and Mesoamerican genepools. Independent domestications in each genepool gave rise to the Andean and Mesoamerican cultivars [18–20]. On the other hand, wild lima bean (*P. lunatus* L.) diverged from common bean, after which natural spread also led to a strong genepool structure, with two Andean and two Mesoamerican genepools. Further independent domestications happened in one Andean and one Mesoamerican genepools [21].

With this in mind, in this chapter we discuss how the recurrent phylogeographic splits and nested domestication events of common and lima beans help understand whether genomic islands of speciation in *Phaseolus* species are more prone to harbor within-species divergence due to reduced recombination and increased drift (**Figure 1**). We concretely focus our discussion by asking the following questions:

1. Are between-species F_{ST} outliers recovered in within-species comparisons?
2. Is there any parallelism in the within-species divergence F_{ST} profiles?
3. Are low-recombining regions (i.e., centromeres) more prone to exhibit divergence across repeated and hierarchically nested scales of divergence?

If there were some parallelisms in the genetic adaptations to the Mesoamerican and Andean environments or in the genetic consequences of the domestication syndromes, then there would be matching signals of differentiation in the within-species between-genepool divergence F_{ST} profiles and in the within-genepool wild-cultivated divergence F_{ST} profiles, respectively. These patterns of repeatability would not be observed if between-genepool and wild-cultivated divergence outliers were due to genetic drift [5], if selection pressures were different [22] or if equivalent selective forces did not act on the same shared variation [6, 23].

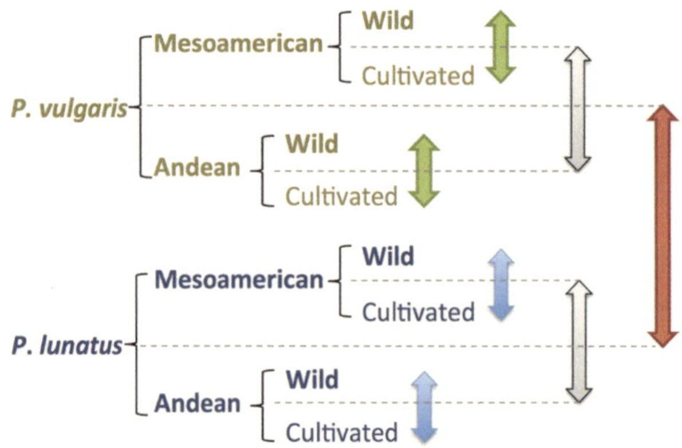

Figure 1. Schematic representation of a sampling across hierarchically nested sampling levels of divergence.

Yet, genomic constrains, rather than true signals of convergent adaptation and domestication, could still be the reason for these parallelisms. If genomic features were indeed constraining divergence, then genomic islands of differentiation would coincide with low-recombining regions regardless the nature and the scale of divergence.

3. Evidence that genomic features constrain divergence across scales

By looking at the genomic diversity patterns in common and lima beans [24–30], there is evidence that differentiation across repeated and hierarchically nested levels of divergence always co-occurs with regions of low SNP density (**Figure 2**). Increased lineage sorting, and consequently rapid differentiation, is a common phenomenon in low-recombining regions because of linked selection and a reduction in the effective population size [8–10]. Likewise, low-recombining regions also tend to exhibit a decline in diversity due to background selection and, to a lower extent, because of genetic hitchhiking [11]. This can be understood as evidence that regions with low SNP diversity are enriched for contiguous signatures of differentiation between bean species, between genepools, and as part of the multiple domestication syndromes. These concurring signatures could be a by-product of genomic constrains inherent to low-recombining regions.

One of the regions that repeatedly exhibit high differentiation across hierarchically nested levels of divergence in the presence of low SNP density is the centromeric section of chromosome Pv11. The wild-cultivated divergence peak in this chromosome is shared by three domestication syndromes and is located beside the outlier peak detected for all within-species between-genepool comparisons, which in turn coincides with a major between-species peak. In this wide section of chromosome Pv11, there are indications that convergent divergence is consistently correlated with very low SNP density, as expected because of combined effects of linked and background selection in low-recombining regions [8–10, 22]. The observation that genomic constrains are biasing divergence across scales in this section of chromosome Pv11

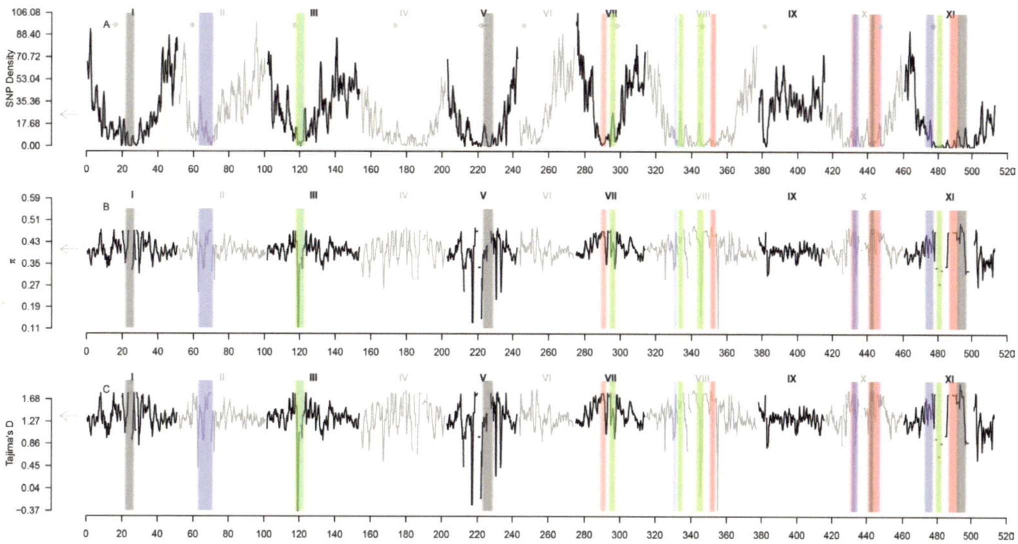

Figure 2. Patterns of genome-wide diversity in common bean and lima beans based on 13,213 GBS-derived SNP markers. A sliding window analysis (window size = 1 × 10^7 bp, step size = 500 kb) was used to compute (**A**) SNP density, (**B**) nucleotide diversity as measured by π, and (**C**) Tajima's D. Vertical translucent boxes highlight the 1 Mb flanking region of each F_{ST}-based outlier window midpoint when F_{ST} was computed as follows: (**red boxes**) between species (*P. lunatus* versus *P. vulgaris*), (**gray boxes**) between genepools (average of four within-species between-genepool comparisons), (**green boxes**) between domestication statuses for *P. vulgaris* (average of two within-genepool wild-cultivated comparisons), and (**blue boxes**) between domestication statuses for *P. lunatus* (average of three within-genepool wild-cultivated comparisons). Results of all windowed analyses are plotted against window midpoints in millions of base pairs (Mb). Black and gray colors highlight different common bean (Pv) chromosomes. Gray arrows on the vertical axes indicate genome-wide averages. Horizontal gray lines with a central-filled gray dot at the top of the figure mark the centromeres [from 20] (figure modified from [16]).

is reinforced by the fact that previous genomic scans did not attribute to this region a consisted outstanding role during the domestication syndromes [20, 21] or in conferring adaptation to different environments and latitudes across the Americas [31]. The only exception is the candidate gene influencing plant size (*Phvul.011G213300*) as part of the Mesoamerican domestication syndrome of common bean [20], but then this pattern has not been consistently reported for the other domestication events as to explain its steady repeatability across hierarchically nested levels of divergence in windows with low SNP density.

Other "hotspots" for spurious divergence due to genomic constrains may be the regions with low SNP density in chromosomes Pv8 and Pv10 that exhibit signatures of between-species divergence as well as repeated between-genepool and within-genepool wild-cultivated divergence (**Figure 2**). The region in chromosome Pv8 was previously reported to be highly divergent during the domestication of the Andean common bean, but then there were not candidate genes in this region associated with that domestication syndrome in particular [20], despite that the same region is known for being involved in plant and seed growth (i.e., *Phvul.008G168000*) during the Mesoamerican domestication of the same species. This paradox may then be a consequence of genomic constrains obscuring genuine anthropic selection and repeatedly forcing divergence in this region. Similarly, the wide divergent region in chromosome Pv10, characterized by two outlier peaks split by a "high valley," actually matches a pericentric inversion between species [32], exemplifying how genomic features inexorably condition differentiation across scales of divergence.

The observation that low-recombining regions are enriched for differentiation across repeated and hierarchically nested levels of divergence in *Phaseolus* beans opposes the profiles of the genome-wide selection scans carried out in common bean. While low-recombining regions are more prone to exhibit signatures of divergence, regions toward the arms of the chromosomes with high SNP density more often harbor adaptive variation [31]. This trend follows expectations because low-recombining regions are more liable to display divergence because of linked selection [11, 33, 34], whereas recombination hotspots usually exhibit higher SNP density and are enriched with functional genes [11, 35]—an already well-described relationship for common bean [36, 37]. Also, adaptive divergent selection usually homogenizes haplotypes within the same niche and fixes polymorphisms in different populations, so that few haplotypes with high frequency remain. This selective process leads to high values of nucleotide diversity and Tajima's D and low values of the Watterson's theta (θ) estimator [38], a tendency that was corroborated in wild common bean when looking for adaptive variants [31] but that was lacking in the present study while retrieving the genomic landscape of divergence between species, genepools, and domestication statuses.

4. Signatures of shared within-species parallel divergence

There is some evidence of some parallelisms in the genetic adaptations to the Mesoamerican and Andean environments in common and lima beans (**Figure 2**). The landscape of genomic adaptation has remained largely unexplored in *Phaseolus* beans. Among the few other studies addressing this question, a panel of wild common bean sampled across the Andean and Mesoamerican ranges revealed that regardless the strength of the bottlenecks [39], the signatures of divergent adaptation are widespread along the genome and coincided with regions of elevated SNP density [31], frequent recombination, and high gene content [36]. However, these surveys have not explicitly addressed the colonization of the Andes by linages coming from Central America and the corresponding change in selection pressures associated with different altitudes, latitudes, and microenvironments. Topographically complex mountainous systems, such as the Andes, harbor an impressive heterogeneity of climates at a small scale [40–43]. The ridges and valleys constitute physical barriers that limit dispersal and cause local variation in rainfall, resulting in genetic isolation and variation in habitats. Both processes have likely speeded up the evolution of high species diversity in this region [44–48]. Yet, the relative effects of geographic isolation [49–51], environmental variation at a small scale [52–58], and their potential interactions across genepools remain poorly understood in wild beans. Therefore, characterizing the genomic consequences associated with the colonization of heterogeneous environments may ultimately disclose further cases of genetic parallelism in the adaptation of beans.

The genomic consequences of multiple domestication events are also moderately recurrent as revealed by our survey. From the twelve regions putatively differentiated as the result of the domestication syndrome, only five (42%) appear in more than one comparison but none appears in all. Two peaks in chromosome Pv3 and Pv10 are repeated across three different comparisons of all five profiles of the domestication syndromes. At least the region in chromosome Pv3 has been reported to be involved in the vernalization pathway (i.e., *Phvul.003G033400*) as part of the Mesoamerican domestication of common bean [20]. Two other divergence peaks in chromosomes Pv8 and Pv11 are consistent across all three genomic profiles of the Mesoamerican domestication syndrome. The region in chromosome Pv8 is known for being related with the encoding of the nitrate reductase (i.e., *Phvul.008G168000*), a critical element for plant and seed growth, during the Mesoamerican domestication of common bean [20]. Also as part of this domestication event, the region in chromosome Pv8 is associated with increased plant size through the ubiquitin ligase degradation pathway (i.e., *Phvul.011G213300*) that controls flower and stem size [20]. More loosely, a peak at chromosome Pv2 in the Mesoamerican common bean domestication F_{ST} profile is recovered in the profiles of all three lima bean domestications. This region has been linked with the domestication syndrome of lima bean since it is involved in the regulation of seed germination (i.e., *Phvul.002G033500*) and leaf size (i.e., *Phvul.002G041800*) and is enriched by inflated linkage disequilibrium scores [21]. Although scattered, some of these few regions may reveal true parallelisms in the domestication syndromes, whereas others may still be constrained by genomic features.

Also striking is the rarity of regions putatively involved in domestication and shared by several domestication events. This trend, mostly expected for quantitative traits with complex genetic architectures [59–61], had already been noticed for the common bean [20]—potentially applying for lima bean as well [21], and so does not necessarily speak for a prevalent role of drift. Since divergence in the lack of repeatability is a liable result of lineage sorting, caution must be undertaken while interpreting these signals. Singularities may result from different adaptive pressures across the Americas unique to each species, distinctive adaptation to the Mesoamerican microenvironments, dissimilar selection as part of each domestication event [22], equivalent selective forces acting on different genetic variants [6, 23], or genetic drift [5]. Discerning among these causes requires further genotyping in an extended panel specifically addressing each comparison. At least for the divergence peak at chromosome Pv7 in the wild-cultivated Mesoamerican common bean comparison, other drivers besides the domestication itself are an unlikely reason for divergence because a wide region in chromosome Pv7 region is known for being associated with increased seed weight (i.e., *Phvul.007G094299-Phvul.007G.99700*) during the Mesoamerican domestication of common bean [20], as well as with flowering regulation (i.e., *Phvul.007G096500* and *Phvul.007065600*) as part of the domestication of lima bean [21] and both common bean genepools [20].

5. Take-home message

Genomic islands of speciation are not necessarily more prone to harbor within-species divergence, yet subjacent genomic constrains could still be shaping parallel divergence at broader genomic scales. With that in mind, we first discussed how genomic features and linked selection could enhance convergent differentiation in low-recombining regions. Later, we reviewed cases of moderate repeatability in the genomic consequences of multiple adaptation and domestication events. This chapter emphasizes that differentiation across repeated and hierarchically nested levels of divergence co-occurs with regions of low SNP density, and these concurring signatures may be a by-product of genomic constrains inherent to low-recombining regions. We advise a more systematic use of repeated and hierarchically nested samplings in order to improve our understanding of the underlying causes of the genomic landscape of divergence. Because certain regions are more prone to accumulate islands of divergence as the result of genomic constrains, we advocate that studies of genomic divergence should consider more systematically a dual-purpose sampling, such as the one we described in the first section. In the first place, using replicated populations under presumably similar selection pressures helps accounting for lineage sorting and characterizing the nature of the selected variants, i.e., novel versus standing [6]. Second, a hierarchically nested sampling across various levels of divergence allows for further assessments on the processes, which like genomic constrains, may give rise to parallel divergence patterns [2–4, 62]. Finally, some of these examinations must be verified with genomic features and estimates of the recombination rate [63–65]. We foresee that as the evidence of pervasive genomic constrains shaping genomic differentiation across species and at countless scales of divergence accumulates, replicated samplings of contrasting populations in a hierarchically nested framework of divergence will become indispensable.

In the long run, we are looking forward to see more coherent and systematic samplings of replicated contrasting populations across hierarchically nested levels of divergence in of genomic divergence has always been challenging, but the field is now moving forward toward a more cohesive framework. New ways [66, 67] to characterize obscuring genomic features promise aiding our understanding on how the genomic landscape of divergence is shaped.

Among the five domesticated species in the *Phaseolus* genus, common and lima beans are the only ones exhibiting range expansions toward South American and multiple domestications [14]. However, exploring the landscape of divergence in other domesticated *Phaseolus* species is equally insightful because of their overlapping distribution ranges, nested phylogenetic relationships, and divergent adaptations. For instance, year (*P. dumosus*) and runner (*P. coccineus*) beans are Mesoamerican and well adapted to humid habitats, which makes them a potential source of resistance to biotic stresses. On the other hand, tepary bean (*P. acutifolius*) is also Mesoamerican but is well known for growing in desert and semiarid environments, which makes it a likely source of tolerance to abiotic stresses. These species also possess well-established genomic resources [68] that could speed up newer genome-wide comparisons. *Phaseolus* species that never underwent domestication are also abundant (ca. 70) and could enrich our understanding of genomic divergence in this intricate complex. Considering the *Phaseolus* species complex as a whole will ultimately reinforce beans as a model for understanding speciation, adaptation, and crop evolution [14, 15, 69–72].

Acknowledgements

Some of the ideas presented in this chapter were refined thanks to the comments from A. Caro, I. Cerón, C. Jiggins, D. Londoño, P. Reyes, C. Salazar, J.J. Wiens, and R. Yockteng during the VI Symposium of the Colombian Society for Evolution held in Cali (Colombia) on August 2017. This chapter was funded by a Colciencias (Colombia) grant awarded to MC under contract number FP44842-009-2015 and project code 1101-658- 42502, by the grant 3404 from Fundación para la Promoción de la Investigación y la Tecnología del Banco de la República de Colombia to MC, and by the Lundell and Tullberg (Sweden) grants to AC. The Geneco mobility fund from Lund University is thanked for subsidizing the meeting between AC and MB in the spring of 2015 at Nashville (TN, USA). AC's writing time was sponsored by the grants 4.1-2016-00418 from Vetenskapsrådet (VR) and BS2017-0036 from Kungliga Vetenskapsakademien (KVA). MB received support from the Evans-Allen fund of the US Department of Agriculture. The editorial fund from the Colombian Corporation for Agricultural Research is acknowledged for financing this publication.

Author details

Andrés J. Cortés[1,2]*, Paola Hurtado[3,4], Mathew W. Blair[5] and María I. Chacón-Sánchez[3]

*Address all correspondence to: acortes@agrosavia.co

1 Corporación Colombiana de Investigación Agropecuaria (Agrosavia), Rionegro, Colombia

2 Universidad Nacional de Colombia - Medellín, Facultad de Ciencias Agrarias - Departamento de Ciencias Forestales, Medellín, Colombia

3 Universidad Nacional de Colombia -Facultad de Ciencias Agrarias - Departamento de Agronomía, Bogotá, Colombia

4 Department of Plant Sciences, University of California, Davis, California, USA

5 Department of Agricultural and Environmental Science, Tennessee State University, Nashville, USA

References

[1] Nei M. The neutral theory of molecular evolution in the genomic era. Annual Review of Genomics and Human Genetics. 2010;**11**:265-289

[2] Barrett RD, Schluter D. Adaptation from standing genetic variation. Trends in Ecology & Evolution. 2008;**23**(1):38-44

[3] Pritchard JK, Pickrell JK, Coop G. The genetics of human adaptation: Hard sweeps, soft sweeps, and polygenic adaptation. Current Biology. 2010;**20**(4):R208-R215

[4] Hermisson J, Pennings PS. Soft sweeps: Molecular population genetics of adaptation from standing genetic variation. Genetics. 2005;**169**(4):2335-2352

[5] Lotterhos KE, Whitlock MC. The relative power of genome scans to detect local adaptation depends on sampling design and statistical method. Molecular Ecology. 2015;**24**(5): 1031-1046

[6] Roesti M et al. The genomic signature of parallel adaptation from shared genetic variation. Molecular Ecology. 2014;**23**(16):3944-3956

[7] Strasburg JL et al. What can patterns of differentiation across plant genomes tell us about adaptation and speciation? Philosophical Transactions of the Royal Society, B: Biological Sciences. 2011;**367**(1587):364-373

[8] Wolf JB, Ellegren H. Making sense of genomic islands of differentiation in light of speciation. Nature Reviews. Genetics. 2017;**18**(2):87-100

[9] Jones FC et al. The genomic basis of adaptive evolution in three spine sticklebacks. Nature. 2012;**484**(7392):55-61

[10] Zhou L, Bawa R, Holliday JA. Exome resequencing reveals signatures of demographic and adaptive processes across the genome and range of black cottonwood (*Populus trichocarpa*). Molecular Ecology. 2014;**23**(10):2486-2499

[11] Ellegren H, Galtier N. Determinants of genetic diversity. Nature Reviews. Genetics. 2016;**17**(7):422-433

[12] Kelleher CT, Wilkin J, Zhuang J, Cortés AJ, Quintero ÁLP, Gallagher TF, et al. SNP discovery, gene diversity, and linkage disequilibrium in wild populations of Populus tremuloides. Tree Genetics & Genomes. 2012;**8**(4):821-829. DOI: 10.1007/s11295-012-0467-x

[13] Cortés AJ. On The Big Challenges of a Small Shrub: Ecological Genetics of Salix herbacea L. Uppsala: Acta Universitatis Upsaliensis; 2015

[14] Bitocchi E et al. Beans (*Phaseolus* ssp.) as a model for understanding crop evolution. Frontiers in Plant Science. 2017;**8**:722

[15] Broughton WJ et al. Beans (*Phaseolus* spp.)—model food legumes. Plant and Soil. 2003; **252**(1):55-128

[16] Cortés AJ et al. Does the genomic landscape of species divergence in *Phaseolus* beans reveal parallel signatures of adaptation and domestication? Frontiers in Plant Science. In revision

[17] Rendon-Anaya M et al. Genomic history of the origin and domestication of common bean unveils its closest sister species. Genome Biology. 2017;**18**(1):60

[18] Gepts P, Debouck D. Origin, domestication and evolution of the common bean (*Phaseolus vulgaris* L.). In: Van Shoonhoven A, Voysest O, editors. Common Beans: Research for Crop Improvement. Wallingford: Commonwealth Agricultural Bureau; 1991. pp. 7-53

[19] Kwak M, Gepts P. Structure of genetic diversity in the two major gene pools of common bean (*Phaseolus vulgaris* L., *Fabaceae*). Theoretical and Applied Genetics. 2009;**118**(5): 979-992

[20] Schmutz J et al. A reference genome for common bean and genome-wide analysis of dual domestications. Nature Genetics. 2014;**46**(7):707-713

[21] Chacon-Sanchez MI, Martinez-Castillo J. Testing domestication scenarios of lima bean (*Phaseolus lunatus* L.) in Mesoamerica: Insights from genome-wide genetic markers. Frontiers in Plant Science. 2017;**8**:1551

[22] Ravinet M et al. Interpreting the genomic landscape of speciation: A road map for finding barriers to gene flow. Journal of Evolutionary Biology. 2017;**30**(8):1450-1477

[23] Ravinet M et al. Shared and nonshared genomic divergence in parallel ecotypes of *Littorina saxatilis* at a local scale. Molecular Ecology. 2016;**25**(1):287-305

[24] Cortés AJ, Chavarro MC, Blair MW. SNP marker diversity in common bean (*Phaseolus vulgaris* L.). Theoretical and Applied Genetics. 2011;**123**(5):827-845

[25] Blair MW, Soler A, Cortés AJ. Diversification and population structure in common beans (*Phaseolus vulgaris* L.). Plos One. 2012;**7**(11):e49488

[26] Cortés AJ, Chavarro MC, Madriñán S, This D, Blair MW. Molecular ecology and selection in the drought-related Asr gene polymorphisms in wild and cultivated common bean (*Phaseolus vulgaris* L.). BMC Genetics

[27] Cortés AJ, This D, Chavarro C, Madriñán S, Blair MW. Nucleotide diversity patterns at the drought-related DREB2 encoding genes in wild and cultivated common bean (*Phaseolus vulgaris* L.). Theoretical and Applied Genetics. 2012;**125**(5):1069-1085

[28] Galeano CH, Cortés AJ, Fernandez AC, Soler A, Franco-Herrera N, Makunde G, et al. Gene-based single nucleotide polymorphism markers for genetic and association mapping in common bean. BMC Genetics. 2012;**13**(1):48

[29] Blair MW, Cortés AJ, Penmetsa RV, Farmer A, Carrasquilla-Garcia N, Cook DR. A high-throughput SNP marker system for parental polymorphism screening, and diversity analysis in common bean (*Phaseolus vulgaris* L.). Theoretical and Applied Genetics. 2013;**126**(2):535-548

[30] Blair MW, Cortés AJ, This D. Identification of an ERECTA gene and its drought adaptation associations with wild and cultivated common bean. Plant Science. 2016;**242**:250-9

[31] Cortés AJ, Blair MW. Genotyping by sequencing and genome—Environment associations in wild common bean predict widespread divergent adaptation to drought. Frontiers in Plant Science. 2018;**9**:128

[32] Bonifácio EM et al. Comparative cytogenetic mapping between the lima bean (*Phaseolus lunatus* L.) and the common bean (*P. vulgaris* L.). Theoretical and Applied Genetics. 2012;**124**(8):1513-1520

[33] Ellegren H, Wolf JBW. Parallelism in genomic landscapes of differentiation, conserved genomic features and the role of linked selection. Journal of Evolutionary Biology. 2017;**30**(8):1516-1518

[34] Ellegren H et al. The genomic landscape of species divergence in *Ficedula* flycatchers. Nature. 2012;**491**(7426):756-760

[35] Rodgers-Melnick E et al. Recombination in diverse maize is stable, predictable, and associated with genetic load. Proceedings of the National Academy of Sciences. 2015;**112**(12): 3823-3828

[36] Blair MW et al. Uneven recombination rate and linkage disequilibrium across a reference SNP map for common bean (*Phaseolus vulgaris* L.). PLoS One. 2018;**13**(3):e0189597

[37] Bhakta MS, Jones VA, Vallejos CE. Punctuated distribution of recombination hotspots and demarcation of pericentromeric regions in *Phaseolus vulgaris* L. PLoS One. 2015;**10**(1): e0116822

[38] Wakeley J. Coalescent Theory: An Introduction. Cambridge: Harvard University; 2008

[39] Ariani A, Berny Mier YTJ, Gepts P. Spatial and temporal scales of range expansion in wild *Phaseolus vulgaris*. Molecular Biology and Evolution. 2018;**35**(1):119-131

[40] Cortés AJ, Blair MW. Naturally available genetic adaptation in common bean and its response to climate change. In: Srinivasarao C, Shanker AK, Shanker C, editors. Climate Resilient Agriculture—Strategies and Perspectives. Rijeka, Croatia: InTech; 2018

[41] Cortés AJ, Wheeler JA. The environmental heterogeneity of mountains at a fine scale in a changing world. In: Hoorn C, Perrigo A, Antonelli A, editors. Mountains, Climate, and Biodiversity. NY: Wiley; 2018

[42] Cortés AJ. Environmental Heterogeneity at a Fine Scale: Ecological and Genetic Implications in a Changing World. Saarbrücken, Germany: LAP Lambert Academic Publishing; 2016. 60 p

[43] Cortés AJ. Local scale genetic diversity and its role in coping with changing climate. In: Bitz L, editor. Genetic Diversity. Rijeka, Croatia: InTech; 2017. p. 140 pp

[44] Hoorn C et al. Amazonia through time: Andean uplift, climate change, landscape evolution, and biodiversity. Science. 2010;**330**(6006):927-931

[45] Antonelli A et al. Tracing the impact of the Andean uplift on Neotropical plant evolution. Proceedings of the National Academy of Sciences. 2009;**106**(24):9749-9754

[46] Madriñán S, Cortés AJ, Richardson JE. Páramo is the world's fastest evolving and coolest biodiversity hotspot. Frontiers in Genetics. 2013;**4**:192

[47] Zizka A, Antonelli A. Mountains of diversity. Nature. 2018;**555**:173-174

[48] Arnegard ME et al. Genetics of ecological divergence during speciation. Nature. 2014; **511**(7509):307-311

[49] Cortés AJ, Waeber S, Lexer C, Sedlacek J, Wheeler JA, van Kleunen M, et al. Small-scale patterns in snowmelt timing affect gene flow and the distribution of genetic diversity in the alpine dwarf shrub *Salix herbacea*. Heredity. 2014;**113**:233-239

[50] Cortés AJ, Liu X, Sedlacek J, Wheeler JA, Lexer C, Karrenberg S. Maintenance of female-bias in a polygenic sex determination system is consistent with genomic conflict. On The Big Challenges of a Small Shrub: Ecological Genetics of Salix herbacea L. Uppsala: Acta Universitatis Upsaliensis; 2015

[51] Cortés AJ, Wheeler JA, Sedlacek J, Lexer C, Karrenberg S. Genome-wide patterns of microhabitat-driven divergence in the alpine dwarf shrub Salix herbacea L. On The Big Challenges of a Small Shrub: Ecological Genetics of Salix herbacea L. Uppsala: Acta Universitatis Upsaliensis; 2015

[52] Sedlacek J, Bossdorf O, Cortés AJ, Wheeler JA, van-Kleunen M. What role do plant-soil interactions play in the habitat suitability and potential range expansion of the alpine dwarf shrub *Salix herbacea*? Basic and Applied Ecology. 2014;**15**(4):305-315

[53] Wheeler JA, Hoch G, Cortés AJ, Sedlacek J, Wipf S, Rixen C. Increased spring freezing vulnerability for alpine shrubs under early snowmelt. Oecologia. 2014;**175**(1):219-229. DOI: https://doi.org/10.1007/s00442-013-2872-8

[54] Sedlacek J, Wheeler JA, Cortés AJ, Bossdorf O, Hoch G, Lexer C, et al. The response of the alpine dwarf shrub salix herbacea to altered snowmelt timing: Lessons from a multi-site transplant experiment. PloS One. 2015;**10**(4):e0122395

[55] Wheeler JA, Schnider F, Sedlacek J, Cortés AJ, Wipf S, Hoch G, et al. With a little help from my friends: community facilitation increases performance in the dwarf shrub Salix herbacea. Basic and Applied Ecology. 2015;**16**:202-229

[56] Little CJ, Wheeler JA, Sedlacek J, Cortés AJ, Rixen C. Small-scale drivers: the importance of nutrient availability and snowmelt timing on performance of the alpine shrub Salix herbacea. Oecologia. 2016;**180**(4):1015-1024

[57] Sedlacek J, Cortés AJ, Wheeler JA, Bossdorf O, Hoch G, Klapste J, et al. Evolutionary potential in the Alpine: Trait heritabilities and performance variation of the dwarf willow Salix herbacea from different elevations and microhabitats. Ecology and Evolution. 2016;**6**(12):3940-3952

[58] Wheeler JA, Cortés AJ, Sedlacek J, Karrenberg S, van Kleunen M, Wipf S, et al. The snow and the willows: Accelerated spring snowmelt reduces performance in the low-lying alpine shrub Salix herbacea. Journal of Ecology. 2016;**104**(4):1041-1050

[59] Meyer RS, DuVal AE, Jensen HR. Patterns and processes in crop domestication: An historical review and quantitative analysis of 203 global food crops. The New Phytologist. 2012;**196**(1):29-48

[60] Blair MW, Iriarte G, Beebe S. QTL analysis of yield traits in an advanced backcross population derived from a cultivated Andean x wild common bean (*Phaseolus vulgaris* L.) cross. Theoretical and Applied Genetics. 2006;**112**(6):1149-1163

[61] Cortés AJ, Monserrate F, Ramírez-Villegas J, Madriñán S, Blair MW. Drought tolerance in wild plant populations: The case of common beans (Phaseolus vulgaris L.). Plos One. 2013;**8**(5):e62898

[62] Pereira RJ et al. Transcriptome-wide patterns of divergence during allopatric evolution. Molecular Ecology. 2016;**25**(7):1478-1493

[63] Maynard Smith J, Haigh J. The hitch-hiking effect of a favourable gene. Genetics Research. 1974;**23**(01):23-35

[64] Nielsen R. Molecular signatures of natural selection. Annual Review of Genetics. 2005;**39**: 197-218

[65] Storz JF. Using genome scans of DNA polymorphism to infer adaptive population divergence. Molecular Ecology. 2005;**14**:671-688

[66] Cortés AJ. On how role versatility boosts an STI. Journal of Theoretical Biology. 2018;**440**: 66-69

[67] Cortés AJ. Prevalence in MSM Is Enhanced by Role Versatility. In: Mazari A, editor. Big Data Analytics in HIV/AIDS Research. Hershey, USA: IGI Global; 2018

[68] Guerra-Garcia A et al. Domestication genomics of the open-pollinated scarlet runner bean (*Phaseolus coccineus* L.). Frontiers in Plant Science. 2017;**8**:1891

[69] Cortés AJ. On the origin of the common bean (*Phaseolus vulgaris* L.). American Journal of Plant Sciences. 2013;**4**(10):1998-2000

[70] Irwin DE et al. Recurrent selection explains parallel evolution of genomic regions of high relative but low absolute differentiation in a ring species. Molecular Ecology. 2016;**25**(18):4488-4507

[71] Nachman MW, Payseur BA. Recombination rate variation and speciation: Theoretical predictions and empirical results from rabbits and mice. Philosophical Transactions of the Royal Society of London. Series B, Biological Sciences. 2012;**367**(1587):409-421

[72] Cruickshank TE, Hahn MW. Reanalysis suggests that genomic islands of speciation are due to reduced diversity, not reduced gene flow. Molecular Ecology. 2014;**23**(13):3133-3157

Chapter 4

Induced Mutation: Creating Genetic Diversity in Plants

Kamile Ulukapi and Ayse Gul Nasircilar

Additional information is available at the end of the chapter

http://dx.doi.org/10.5772/intechopen.81296

Abstract

Genetic diversity is the variation occurred in genetic information, which depends on frequency and diversity of alleles among individuals within a population or a species. This phenomenon, which is also a part of the evolution process, allows the organisms to adapt to changing conditions and to survive. Populations with high allelic variability are more easily adaptable to changing environmental conditions. However, nowadays, constant use of populations with certain characters in the plant breeding and the uniformity of consumer demands are among one of the causes of genetic erosion. Loss of genetic diversity within a species can lead to loss of useful properties for human beings. If stress conditions such as disease or drought occur, the ability of a population to survive by adapting to this new condition is dependent on the presence of individuals carrying gene alleles that need to adapt to these conditions.

Keywords: gamma rays, genetic diversity, induced mutation, plant

1. Introduction

The first information on the importance of genetic diversity begins with Darwin, which emphasizes the importance of variations in the process of adaptation to natural habitat. In the process of surviving species that can adapt to changing environmental conditions and the disappearance of others, the factors that caused them were researched and the definition of alleles was first used by Mendel, who is considered the father of heredity. Alleles do not only create the source of similarities and differences between progenies, but also makes it possible for species with high genetic diversity to continue their evolutionary process by adapting to different conditions. A biochemical approach to genetic diversity was presented by the introduction of isoenzyme techniques in the mid-1960s. In the following years, the discovery of the structure of DNA and its DNA-based examination of diversity has enabled scientists to reach accurate information about genetic diversity, gene flow, and the origins of species. According to the "Neutral Theory of Molecular Evolution" first introduced by Motoo Kimura in 1968, many of the evolutionary changes arise from the genetic drift of the neutral mutant alleles. Greater than 100 bp insertion/deletion is rapidly removed by natural selection that has caused a great change and damage to the DNA. According to the neutral hypothesis, mutations that occur with smaller changes are harmless mutations. They are protected in the evolutionary process by contributing to the formation of genetic diversity. Most of the polymorphisms are kept in a population by mutation and random genetic drift [1–7].

The genetic difference between individual of a species is the basis for evolution and adaptation. If all of the individuals were genetically identical, species could not be survived under changing environmental conditions such as drought and salinity. It is known that there is a positive correlation between fertility and viability with population size and genetic variation. For this reason, regardless of whether the existing genetic diversity is beneficial, it should be considered that it poses a potential against rapidly changing environmental conditions [8, 9]. Loss of genetic diversity will adversely affect the ability to deal with environmental change in the evolutionary process. It is anticipated that, particularly small and isolated populations will be more affected by this loss [4]. Although loss of genetic diversity has been thought to be a threat to rare species for a long time, it has also been found to be effective on widespread species with large populations [4, 10, 11]. Furthermore, the fertility and viability rate of populations with reduced genetic variability and loss of valuable alleles, especially as a result of self-destruction, is diminishing. Random genetic drift changes the frequency of alleles between generations. The frequency of some of the alleles may decrease while the other allele may become more common. Naturally occurring mutations also support this process. The genetic variation among individuals is the basis for the evolutionary change of species, populations, and progeny. Evolutionary change requires modifications of genes or gene combinations. The functions of the existing alleles are altered by mutation or recombination. Appropriate mutations can help the organism better use the current environment. Alleles that are previously neutral or have little effect on the reproduction success of individuals may suddenly become important. Obtaining useful genetic variations through natural mutations is a very slow process. In the future, it is impossible to predict which alleles will be necessary for a survival. For this reason, populations and species with high genetic variation are more likely to have alleles that may be necessary for adaptation to changing conditions. Generally, natural populations have the variation to be expressed in the case of the change of selection pressure. The basis of plant breeding is also based on this genetic variation. By artificially altering the selection pressure, some alleles come forward and populations gain new features. Since there is a positive correlation between population size and genetic diversity [12], decreasing population size or limiting gene flow between subpopulations may cause a reduction in genetic diversity [13]. Genetic drift with a positive role in the evolutionary process [3], inbreeding, mating between close relative individuals, reduction of viability and fertility of individuals in the population may cause effects such as the reduction of heterozygote, allelic losses, and some alleles become fixed in the population. As a result, genetic variation will be lost in populations [9, 12, 14]. Many populations are so small that genetic drift, a random process, has an important influence on the number of alleles that will be transferred to future generations. The useful alleles transferred from previous generations may be lost as a result of genetic drift [1]. Two main consequences of genetic drift; (a) different alleles frequencies between generations are irregular and (b) the genetic diversity of the populations is lost. At the beginning, rare alleles will disappear and the mean heterozygote will decrease over time. While genetic diversity within populations decrease, genetic differences between populations may increase or decrease depending on whether the random genetic drift in different populations in the same or opposite direction. This loss will continue until the gene pool is stabilized for an allele or until a balance is established between loss of genetic variation and genetic variation through mutations. The loss of both rare alleles and heterozygotes, which cause the decrease in genetic diversity in the population, will decrease in an inverse proportion to the effective population size (Ne). The population size (Ne), which helps to determine the demographic structure of the population, makes it possible to predict the change in genetic diversity. Population size is an important factor that increases the genetic diversity of individuals' intra-species and between taxa. Besides genetic diversity of non-endangered species is higher than endanger species [6, 15–17]. Only a small amount of gene flow may be sufficient to prevent the loss of genetic diversity in a population. Especially, the accumulation of mutations with small deleterious effects in the alleles may reduce the viability and fertility of the populations. These mutations never reach high frequencies in large populations of sexual reproduction. The selection pressure prevents these harmful alleles to become widespread. However, in isolated small populations, the frequency of these harmful alleles may increase and be fixed by chance, if genetic drift may be stronger than natural selection. When a harmful mutation is stabilized in the genetic construct, it causes the population size to decrease and other harmful mutations to be fixed [18]. While the destructive effect of detrimental mutations affect large populations, thousands of years later, can be seen in minutes in small populations. This event, known as mutation meltdown, reduces the size of the population depending on the accumulation of detrimental mutations. The mutation meltdown, is a genetic problem, especially for small populations of endangered species, occur depending on mutation characteristics, demographic characteristics of the mutation and population, and the relationship between mutations and their adaptation [19, 20]. Species or populations suffering from the genetic bottleneck disappear over time. Adaptation can sometimes be a rapid process involving a single gene. In cases when the positive selection is strong, the best allele is fixed in the relevant locus and the adaptation process is terminated unless a better allele is generated by mutation or a new allele is introduced into the population by gene flow. This does not occur in polygenic characters. Genetic drift has a significant impact on the protection of genetic diversity and the amount of genetic variation among populations. Genetic drift as well as mutations, selection, and gene flow are factors contributing to genetic diversity. Sometimes, however, subspecies can occur as a result of gene flow, and which can adversely affect fertility or cause outbreeding depression [9, 21].

2. Natural mutagenesis

When damage occurs in DNA due to a physical, chemical, or biological agent, molecular systems recognize and repair this damage. When this mechanism is unsuccessful, mutation occurs in the organism [22]. Mutagenesis, a consequence of errors in DNA repair, is a mutation-producing process. Hereditary changes that occur naturally and suddenly, which are not caused by recombination and segregation, are called mutations [23]. Mutations that lead to the formation of new individuals, species, and genera are considered as the most important factors of evolution since they can be transferred to future generations [24]. Mutations that occur in somatic cells are not transferred to future generations, but they are important for vegetative produced species. Mutation-derived individuals are called "mutants" [23]. Natural mutants formed in the evolutionary process are one of the most important factors contributing to the formation of species. The rate of spontaneous mutations in higher plants is quite low (10^{-5}–10^{-8}) [25]. Mutations occur more frequently in some regions of the genome. For example, in almost all organisms, the mutation rate in GC regions is higher than in AT regions [26]. While deleterious or neutral mutations that form a part of the mutations that occur in the natural process may disappear in the evolutionary process, the protected mutants may have desirable agricultural characteristics or easily adapt to changing environmental conditions [25]. The first document relating to mutant selection belongs to the year 300 BC. The description of various wild and cultivated mutants was made by Linnaeus in the second half of the 1700s [23]. Although hereditary variations have been observed and used for thousands of years, the mechanism of heredity was first revealed by Mendel [27]. Johanssen's research on seed index of common beans in 1913 can be considered as the first to prove the presence of natural mutations with small effects. In 1924, Baur emphasized that the accumulation of these small mutations in the genome over the years had an impact on the evolution process [28]. Mutation term was first used by de Vries at the beginning of the 1900s. The first evidence of mutation breeding work, which will gain a new perspective on plant breeding, was obtained in 1927 in *Datura stramonium* by radium ray application [20]. The process that begins with human being's awareness of natural mutation has led to the development of many new varieties with induced mutation.

Mutations can occur in spontaneously or under different influences of various mutagens. For this reason, there are various classifications made by different researchers [2, 29–33]. Yüce et al. [34] classified mutations into two main categories as genomic and plasmon mutations. Genome mutations; (1) "gene mutations" resulting from genetic changes, (2) "Chromosomal mutations" formed by chromosome aberrations or chromosomal changes, (3) "Ploidy mutations" resulting from genome and chromosome number changes, (4) Mutations created by transposition elements, and (5) Mutations resulting from somaclonal and gametoclonal variations were classified as five subgroups. Mutations occurring in the genetic material of mitochondria and plastids in the cytoplasm are classified as "plasmon mutations" in a single heading [34].

Gene mutations are structural gene changes in DNA that occur through different mechanisms such as deletion, insertion, and substitution. Intercalar substances, ultraviolet rays, alkylating compounds, and free radicals cause gene mutations [33, 34].

Chromosome mutations are genetic changes that are generally caused by deletion, duplication, inversion, and translocation mechanisms and are larger than gene mutations [33, 34].

Ploidy mutations are divided into two main groups as polyploidy and aneuploidy. The smallest ploidy level is the haploid in the gametes, containing n chromosomes. Eukaryotes contain diploid (2n) chromosomes in cell nuclei. Cells that contain more than two genomes in the nucleus are called polyploids. Polyploidy are very common in plant kingdom and are naturally seen in important cultivated plants such as wheat, cotton, potato, banana, and coffee. These species, which contain more than two alleles in terms of genetic structure, display a richer genetic variability. Aneuploidy is the number of chromosome changes that occur as an increase or decrease in the number of chromosomes. In this case, individuals with fewer or more chromosomes than normal chromosomes are formed [34, 35].

Transposon elements are the mobile genetic elements found in the genome and cause mutation due to their ability to displace within the genome [36, 37].

Somaclonal and gametoclonal variants, another source of mutations, are genotypic or phenotypic differentiations in somatic or gametic cells, which are formed by hormones used in tissue culture media [34, 38, 39].

Spontaneous mutations caused by disruptions in the functioning of molecular mechanism in the cell, the main source of genetic diversity [40]. In every generation, 10^{-5}–10^{-6} mutation rate per gamet cell occurs. This ratio can vary between genes and even by regions within the genes [41]. Although mutations occur infrequently, when considered as a whole genome, it plays an important role in the change of genetic diversity. Because mutations occur at randomly, and it cannot know which one of the gene copies will mutate in diploid or polyploid organisms [22].

Spontaneous mutations have been the basis for the beginning of agriculture and for humankind to pass on a settled life. Self-changing hereditary features have made the dormancy period reduced in species such as peas, wheat, and barley. In addition, the loss of bitterness was formed in almond, linden, watermelon, potato, eggplant, cabbage, and various hazelnut species. All these developments have made these products suitable for human consumption. Another spontaneous mutation was the formation of parthenocarpy in grape and banana (**Table 1**). Naturally occurring mutations have led humans to work on induced mutations [42].

Mutation that facilitated domestication	Examples of plant
Abolishment of bitterness and toxicity	almonds, lima beans, watermelons, potatoes, egg-plants, cabbages nuts
Abolishment of the need for sexual reproduction (seedlessness or parthenocarpy)	bananas, grapes, oranges, and pineapples
Loss of natural seed dispersal mechanism—shattering of pods and heads	peas, wheat, barley
Loss of the hard seed coat and other germination inhibitors (dormancy)	wheat, barley, peas
Facility for self-compatible hermaphroditism	grapes, papaya, *etc.*

Table 1. Useful properties acquired by spontaneous mutations in evolutionary processes in plants (Table was directly taken from Mba [42]).

3. Induced mutation

When it considered the changing environmental conditions and population growth, it is necessary to increase agricultural products approximately 70% in near future [43]. However, breeding trials for the development of desirable agricultural characteristics cause genetic bottleneck. Genetic resource erosion in plants will also lead to the loss of useful genes that would potentially create for breeding studies [44]. Induced mutations may help regain lost traits due to reasons, such as stress factors in the evolutionary process. These genotypes, exempted by the ethical and legal limitations faced of genetically modified products, can be identified by advanced molecular techniques. Thus, the variation of the mutants with the new phenotypes revealed can provide a different perspective for plant breeding studies [45]. Mba et al. [46], referring to the importance of landrace and wild varieties as important genetic resources in breeding strategies, proposed that artificial mutation of putative parental materials in order to create new alleles controlling the desired characters for the twenty-first century "smart" crop varieties [42].

There is a 125-year history of studies on induced mutation. It was determined that X-ray, alpha, beta, and gamma rays are the source of radiation, with different studies in 1895–1900. In 1897–1908, the first studies were carried out to investigate the effects of radiation on plants in 1901 and 1911, it is proved that mutation was induced by chemicals at the first time. In 1904 and 1905, Hugo de Vries suggested that radiation promoted artificial mutation. In 1910, Thomas Hunt Morgan did his first mutation experiments with *Drosophila melanogaster*. In 1927, Muller proved precisely that the X-rays induced mutation. In 1928, Lewis John Stadler successfully induced mutation in corn and barley using X-rays. In the years 1934–1938, Tollenar improved the first commercial variety of tobacco called "Chlorina," and this variety was released in Indonesia [28]. After these initial developments, the curiosity and research of the scientists against the induced mutation have continued.

Today, there are 3222 commercial mutant varieties according to the IAEA data. The countries where the most mutant species are released are China (810), Japan (481), and India (330). According to this data, the highest mutant cultivation rate is in Asia continent [47]. When the products are examined, the percentage of mutant varieties by mutation breeding are constituted of 49.5% cereal, 21.9% ornamental plants and flowers, 15% legume, 2.4% fruit nuts, 2.4% vegetable crops, 2.3% fiber crop, 2.1% oil crops, 1.2% forage crops, 0.6% root-tuber crops, 0.4% herbs, 0.2% medicinal plants, and 2% other crops [48].

Mutations can be induced by biological, chemical, or physical factors as well as spontaneous [40]. In breeding studies, physical and chemical mutagens are generally preferred, but mutations also can be generated by biological agents such as viruses and bacteria. While X-rays, γ-rays, fast neutrons, ultraviolet (UV) rays, beta particles, alpha particles, protons, and ion beams are used as physical mutagens, as chemical mutagens; alkylating agents, azide, hydroxylamine, antibiotics, nitroso compounds, acridines, and base analogs are used for mutagenesis [32, 47]. However, "insertional mutagenesis" and "site-directed mutagenesis" methods, which give more precise results in parallel with progress in genome and sequencing studies, are predicted to be used more widely in future mutation breeding studies [49, 50]. Mutant plants were mostly developed using physical mutagens. Physical mutagens are preferred to chemical mutagens for reasons such as ease of use, low cost, and nontoxicity. In particular, gamma and X-rays are the most commonly used mutagens. About 64% of the mutant plants obtained with the physical mutagens were improved using gamma rays [51–53].

Different plant parts such as seed, meristem, callus, and anther can be used in induced mutation studies. Mutation studies begin with the initial phase, called M_0, where different plant materials are used. Each generation of the mutation continues in the form of M_1, M_2, M_3,.... When a seed is used as the starting material, homohistont generation is obtained at the M_2 stage, while the number of cycles may increase in mutagenesis using vegetative tissues. Screening and selection starts with the first homohistont generations. Once these stages are completed, experiments are carried out to release the mutants obtained, or they can be used as parents in breeding programs [23].

4. Use of induced mutations in the enrichment of plant genetic resources

Induced mutation studies were initially conducted only in field conditions. Tissue culture studies, began with cell culture at the beginning of the twentieth century, have become widespread parallel to the development of technology and have enabled the rapid and disease free reproduction of many plant species. *In vitro* mutation may be preferred for tolerance selection, especially for stress and diseases, because of the shortening of the selection period, being economical and the need for small areas in the mutation studies. Generally, plants are transferred to field conditions after the selection made in step M_1V_3 *in vitro* and in this way provides the researcher time and labor savings.

In vitro and *in vivo* conditions, different plant explants and different doses of the mutagen to be administered have used to create variations in plant species. High dose applications will increase mutation frequency in induced mutation studies to create genetic diversity in plants. In this case, however, the percentage of survival of the plants will either be too low or the plant will die [54, 55]. For this reason, appropriate dose determination is required for each mutagen and plant species. Appropriate dose determinations are made on the basis of M_1 plants survival rate and vegetative growth, primarily shoot height [56]. A sample graph, appropriate doses calculated according to survival rate of some common bean cultivars by gamma irradiation are given in **Figure 1** [57].

Numerous studies have been carried out for enriching genetic variation by plant-induced mutation and using this variation to the benefit of humankind (**Figure 2**). Some of these studies were conducted in order to create polymorphism by removing genetic bottleneck. In this way, new hybrid groups can be formed. Genetic diversity is also important for breeding programs and the sustainable use of genetic resources [58]. Genetic variation obtained from induced mutation has contributed to modern plant breeding. Studies conducted by mutation breeding can be summarized under the titles of biotic and abiotic stress tolerance, improving plant nutritional properties, and increasing polymorphism. Barakat and El-Sammak [59] in *Gypsophila paniculata* L., Kaul et al. [60] and Barakat et al. [61] in chrysanthemum obtained mutant

Figure 1. Stages of mutation breeding (Figure was directly taken from Jankowicz-Cieslak et al. [27]).

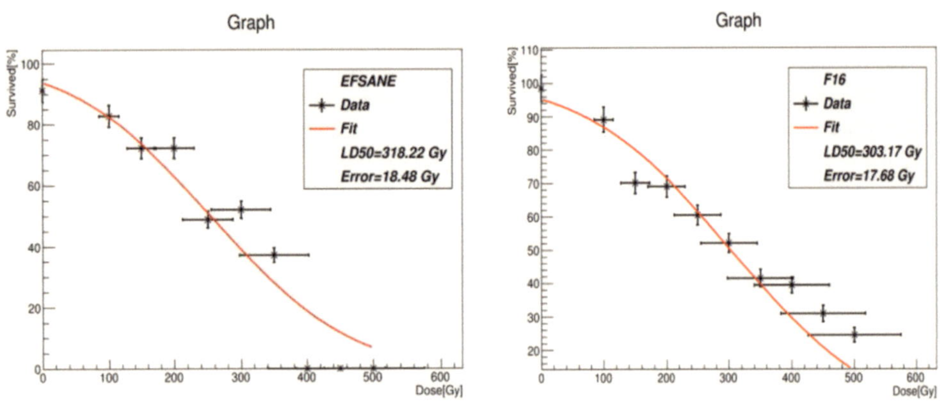

Figure 2. LD50 values of two common bean cultivars (figure was directly taken from Ulukapi and Ozmen [57]).

plants and they identified mutants' genetic similarity with molecular markers (from 0.59 to 0.97, 0.06 to 0.79, and 0.43 to 0.95, respectively). About 83% polymorphism was detected in the chrysanthemum as a result of gamma-induced mutation. It has been reported that the 30 Gy gamma dose is the most effective dose for in vitro genetic variation [62]. Wu et al. [63] have developed resistance at varying frequencies to blast, bacterial blight, and tungoric disease using both chemical (diepoxybutane and ethylmethanesulfonate) and physical (fast neutron and gamma ray) mutagens in rice. The semidwarf rice mutant "Calrose 76" released in California and the short height mutant rice called "Basmati 370" in Pakistan were improved

[64]. EMS is the most preferred chemical mutagen. It causes a single nucleotide polymorphism (SNP). In this way, even a single change in the genomic coding sequences will change the expression of the gene, causing changes in transcription and translation products [65]. Till et al. [66] developed two mutant rice populations using ethyl methanesulphonate (EMS) and a combination of sodium azide plus methyl-nitrosourea (Az-MNU). The investigators have screened target genes and identified 30 nucleotides changes in Az-MNU population and 27 nucleotides in the EMS population. In a study using EMS as a chemical mutagen, four populations with different mutation densities were developed on soybean. The results are described by Targeting Induced Local Lesions IN Genomes (TILLING) [64]. Minoia et al. [68] obtained a new mutant collection at domestication by EMS. All of the genome scans identified 66 nucleotide substitutions and reported two different mutation intensities. In Barley, two gene-induced mutations were generated using EMS and the results were confirmed by sequence analysis [69]. In another study conducted at barley, 63 androgenic doubled-haploid mutants were obtained by sodium azide application during anther formation in vitro [70]. In the study by Kim et al. [71], homologous mutant lines were developed resistant to 5-methyltryptophan (5MT) in rice. In a similar study in rice, mutants which were resistant to 5-methyltryptophan (5MT) were also developed, and the amount of both protein and nine free essential amino acids increased significantly from the original variety [72]. In another study on rice, a new mutant genotype with high tocopherol content was obtained in in vitro mutagenesis with gamma irradiation. Mutant individuals were found to have higher seed viability than the control group and seedling growth was faster in the early growth phase [73]. Induced mutation treatment resulted in acidity and drought tolerance in lentil and rice [74–76]. Again, in a study of rice, salt tolerant varieties were obtained by mutation induction [77]. As seen in all of these studies, many mutagenesis applications have been made in order to improve plant characteristics and to create genetic diversity, and successful results have been achieved. Some methods are used to detect the regions of mutation and density. Methods such as conformation-sensitive capillary electrophoresis (CSCE), single-strand conformation polymorphism (SSCP), and denaturing high performance liquid chromatography (dHPLC) are used to determine variations in plant genes. In addition to these methods, TILLING and High-resolution melting (HRM) are used to determine the induced polymorphism [25, 65, 67].

5. Conclusion

Mutations naturally occurring in the evolutionary process led to the formation of new genotypes. Mutations that occur in nature spontaneously have been modeled for humankind in order to increase the genetic diversity, which is narrowed as a result of natural selection and classical breeding studies. Thus, induced mutation studies have begun. Mutation induction studies, which provide new alleles to the genome by different methods, contribute to the increase of genetic diversity. The increase in genetic variation will increase the chance of survival of species in changing biotic and abiotic conditions. Genetic diversity is not only important for the continuity of species, but also improves the quality criteria of plants, which are raw materials of many industrial products such as food, pharmaceutical, textiles, etc., in these sectors.

Conflict of interest

The authors declare that they have no conflict of interest.

Author details

Kamile Ulukapi[1]* and Ayse Gul Nasircilar[2]

*Address all correspondence to: kamileonal@akdeniz.edu.tr

1 Vocational School of Technical Sciences, Akdeniz University, Antalya, Turkey

2 Department of Mathematics and Science Education, Faculty of Education, Akdeniz University, Antalya, Turkey

References

[1] Arslan RC, Penke L. Evolutionary genetics. In: David M. Buss, editors. Handbook of Evolutionary Psychology. Vol. 2. New Jersey. 2015

[2] Auerbach C. Mutation Research: Problems, Results and Perspectives. London: Chapman and Hall; 1976

[3] Barton N, Partridge L. Limits to natural selection. BioEssays. 2000;**22**(12):1075-1084. DOI: 10.1002/1521-1878

[4] Booy G, Hendriks RJJ, Smulders MJM, Van Groenendael JM, Vosman B. Genetic diversity and the survival of populations. Plant Biology. 2000;**2**(4):379-395. DOI: 10.1055/s-2000-5958

[5] Clegg MT. Plant genetic diversity and the struggel to measure selection. The Journal of Heredity. 1997;**88**(1):1-7. DOI: 10.1093/oxfordjournals.jhered.a023048

[6] Tajima F. Statistical method for testing the neutral mutation hypothesis by DNA polymorphism. Genetics. 1989;**123**(3):585-595

[7] Yamazaki T, Maruyama T. Evidence for the neutral hypothesis of protein polymorphism. Science. 1972;**178**(4056):56-58. DOI: 10.1126/science.178.4056.56

[8] Keller LF, Waller DM. Inbreeding effects in wild populations. Trends in Ecology & Evolution. 2002;**17**(5):230-241. DOI: 10.1016/S0169-5347(02)02489-8

[9] Lundqvist AC, Andersson S, Lönn M. Genetic Variation in Wild Plants and Animals in Sweden: A Review of Case Studies from the Perspective of Conservation Genetics. Swedish Environmental Protection Agency. Report, 5786; 2007

[10] Ellstrand NC, Elam DR. Population genetic consequences of small population size: Implications for plant conservation. Annual Review of Ecology and Systematics. 1993;**24**:217-242. DOI: 10.1146/annurev.es.24.110193.001245

[11] Frankham R. Conservation genetics. Annual Review of Genetics. 1995;**29**:305-327. DOI: 10.1146/annurev.ge.29.120195.001513

[12] Honnay O, Jacquemyn H. Susceptibility of common and rare plant species to the genetic consequences of habitat fragmentation. Conservation Biology. 2007;**21**(3):823-831. DOI: 10.1111/j.1523-1739.2006.00646.x

[13] Leimu R, Mutikainen P, Koricheva J, Fischer M. How general are positive relationships between plant population size, fitness and genetic variation? Journal of Ecology. 2006;**94**: 942-952. DOI: 10.1111/j.1365-2745.2006.01150.x

[14] Young A, Boyle T, Brown T. The population genetic consequences of habitat fragmentation for plants. Trends in Ecology & Evolution. 1996;**11**:413-419. DOI: 10.1016/0169-5347(96)10045-8

[15] Wright S. Evolution in Mendelian populations. Genetics. 1931;**16**:97-159

[16] Kimura M. The Neutral Allele Theory of Molecular Evolution. Cambridge: Cambridge University Press; 1983

[17] Frankham R. Relationship of genetic variation to population size in wildlife. Conservation Biology. 1996;**10**(6):1500-1508. DOI: 10.1046/j.1523-1739.1996.10061500.x

[18] Whitlock MC, Bürger R. Fixation of New Mutations in Small Populations. Interim Report IR-04-064. Austria. 2004. p. 17

[19] Lynch M, Bürger R, Butcher D, Gabriel W. The mutational meltdown in asexual populations. The Journal of Heredity. 1993;**84**(5):339-344. DOI: 10.1093/oxfordjournals.jhered.a111354

[20] Lande R. Mutation and conservation. Conservation Biology. 1995;**9**:782-791. DOI: 10.1046/j.1523-1739.1995.09040782.x

[21] Whiteley AR, Hastings K, Wenburg JK, Frissell CA, Martin JC, Allendorf FW. Genetic variation and effective population size in isolated populations of coastal cutthroat trout. Conservation Genetics. 2010;**11**(5):1929-1943. DOI: 10.1007/s10592-010-0083-y

[22] Futuyma DJ. Evolution. 3rd ed. Sinauer Associates, Inc.; 2013

[23] Foster BP, Shu QY. Plant mutagenesis in crop improvement: Basic terms and applications. In: Shu QY, Forster BP, Nakagawa H, Nakagawa H, editors. Plant Mutation Breeding and Biotechnology. Wallingford: CABI; 2012

[24] Mba C, Afza R, Bado S, Jain SM. Induced mutagenesis in plants using physical and chemical agents. In: Plant Cell Culture: Essential Methods. Vol. 20. Oxford, UK. 2010. pp. 111-130

[25] Jiang SY, Ramachandran S. Natural and artificial mutants as valuable resources for functional genomics and molecular breeding. International Journal of Biological Sciences. 2010;**6**(3):228

[26] Shen JC, Rideout WM, Jones PA. The rate of hydrolytic deamination of 5-methylcytosine in double-stranded DNA. Nucleic Acids Research. 1994;**22**:972-976. DOI: 10.1093/nar/22.6.972

[27] Jankowicz-Cieslak J, Mba C, Till BJ. Mutagenesis for crop breeding and functional genomics. In: Biotechnologies for Plant Mutation Breeding. Cham: Springer; 2017. pp. 3-18

[28] Kharkwal MC. A brief history of plant mutagenesis. In: Shu QY, Forster BP, Nakagawa H, Nakagawa H, editors. Plant Mutation Breeding and Biotechnology. . Wallingford: CABI; 2012

[29] Stubbe H, Genmutation I. Allgemeiner Teil. Handbuch der Ver-Erbungswissenschaft. Band II F. Berlin: Verlag von Gebrüder Bornträger; 1938

[30] Gustafsson Å, Ekberg I. Types of mutations. Manual on Mutation Breeding. Technical Reports No. 119. Venna: IAEA; 1977; pp. 107-123

[31] van Harten AM. Mutation Breeding: Theory and Practical Applications. UK: Cambridge University Press; 1998

[32] Pathirana R. Plant mutation breeding in agriculture. In: Plant Sciences Reviews. UK. 2011. pp. 107-126

[33] Lundqvista U, Franckowiakb JD, Forster BP. Mutation categories. Shu QY, Forster BP, Nakagawa H, Nakagawa H. Plant Mutation Breeding and Biotechnology. CABI. Swedish Environmental Protection Agency. Report, 5786; 2012

[34] Yüce S, Bilgen G, Demir İ. Genetik. Ankara: Nobel Yayınevi; 2010. pp. 218-265

[35] Ramsey J, Schemske DW. Neopolyploidy in flowering plants. Annual Review of Ecology and Systematics. 2002;**33**:589-639. DOI: 10.1146/annurev.ecolsys.33.010802.150437

[36] Bryan G, Garza D, Hartl D. Insertion and excision of the transposable element mariner in Drosophila. Genetics. 1990;**125**(1):103-114

[37] Settles AM. Transposon tagging and reverse genetics. In: Molecular Genetic Approaches to Maize Improvement. Berlin, Heidelberg: Springer; 2009. pp. 143-159

[38] Larkin PJ, Scowcroft WR. Somaclonal variation—a novel source of variability from cell cultures for plant improvement. Theoretical and Applied Genetics. 1981;**60**(4):197-214

[39] George EF. Plant Propagation by Tissue Culture, Part 1: The Technology. England: Exegetics Ltd.; 1993

[40] Sadava DE, Hillis DM, Heller HC, Berenbaum M. Life: The Science of Biology. 10th ed. Sinauer Associates. Sunderland, MA, USA: W.H. Freeman; 2012

[41] Lynch M, Blanchard J, Houle D, Kibota T, Schultz S, Vassilieva L, et al. Spontaneous deleterious mutation. Evolution (N. Y). 1999;**53**:645-663. DOI: 10.1111/j.1558-5646.1999.tb05361.x

[42] Mba C. Induced mutations unleash the potentials of plant genetic resources for food and agriculture. Agronomy. 2013;**3**(1):200-231. DOI: 10.3390/agronomy3010200

[43] Food and Agriculture Organization of the United Nations 2009. How to Feed the World in 2050. Rome, Italy: Food and Agriculture Organization of the United Nations; 2009

[44] Lee CT, Wickneswari R, Clyde MM, Zakri AH. Maintenance of genetic diversity in Parkia speciosa in logged-over forests. Journal of Tropical Forest Science. 2002;**14**(2):163-178

[45] Chaudhary A, Chaudhary S. Quality improvement in plants through induced mutations. In: Tomlekova NB, Kozgar MI, Wani MR, editors. Mutagenesis: Exploring Genetic Diversity of Crops. Wageningen Academic Publishers; 2014. p. 103. DOI: 10.3920/978-90-8686-796-7_11

[46] Mba C, Guimaraes P, Ghosh K. Re-orienting crop improvement for the changing climatic conditions of the 21st century. Agriculture & Food Security. 2012;**1**:7. DOI: 10.1186/2048-7010-1-7

[47] Oladosu Y, Rafii MY, Abdullah N, Hussin G, Ramli A, Rahim HA, et al. Principle and application of plant mutagenesis in crop improvement: A review. Biotechnology and Biotechnological Equipment. 2016;**30**(1):1-16

[48] Joint FAO/IAEA Division of Nuclear Techniques in Food and Agriculture 2018. Mutation Breeding. [Accessed: August 2018]. Available from: http://www-naweb.iaea.org/nafa/pbg/mutation-breeding.html

[49] Sangwana RS, Ochattb S, Nava-Saucedoa JE, Sangwan-Norreela B. T-DNA insertion mutagenesis. In: Shu QY, Forster BP, Nakagawa H, Nakagawa H, editors. Plant Mutation Breeding and Biotechnology. Wallingford: CABI; 2011

[50] Osakabe K, Saika H, Okuzaki A, Toki S. Site-directed mutagenesis in higher plants. In: Shu QY, Forster BP, Nakagawa H, Nakagawa H, editors. Plant Mutation Breeding and Biotechnology. Wallingford: CABI; 2011

[51] Kharkwal MC, Pandey RN, Pawar SE. Mutation breeding for crop improvement. In: Jain HK, Kharkwal MC, editors. Plant Breeding: Mendelian to Molecular Approaches. New Delhi, India: Narosa Publishing House; 2004

[52] Jain SM. Major mutation-assisted plant breeding programs supported by FAO/IAEA. Plant Cell, Tissue and Organ Culture. 2005;**82**:113-123. DOI: 10.1007/s11240-004-7095-6

[53] Jain SM. Mutagenesis in crop improvement under the climate change. Romanian Biotechnological Letters. 2010;**15**(2):88-106

[54] Koornneef M. In: Gilmartin PM, Bowler C, editors. Molecular Plant Biology. Gene Identification: Classical Mutagenesis in Higher Plants. UK: Oxford University Press; 2002. pp. 1-13

[55] Nouri H, Kiani D, Khani MAH. Investigation of mutagenic effects of various doses of gamma ray on seed germination traits of pinto bean cultivar of Khomein. Annals of Biological Research. 2012;**3**(10):4977-4979

[56] Sağel Z, Peşkircioğlu H, Tutluer Mİ. Nükleer Tekniklerin Bitki Islahında Kullanılması, VIII. Ulusal Nükleer Bilimler ve Teknolojileri Kongresi; 15-17 Ekim, 2003. p. 14; Kayseri

[57] Ulukapi K, Ozmen SF. Study of the effect of irradiation (^{60}Co) on M_1 plants of common bean (*Phaseolus vulgaris* L.) cultivars and determined of proper doses for mutation breeding. Journal of Radiation Research and Applied Science. 2018;**11**(2):157-161. DOI: 10.1016/j.jrras.2017.12.004

[58] Mwacharo JM, Otieno CJ, Okeyo MA. Suitability of blood protein polymorphisms in assessing genetic diversity in indigenous sheep in Kenya. In: Applications of Gene-Based Technologies for Improving Animal Production and Health in Developing Countries. Dordrecht: Springer; 2005. pp. 585-591. DOI: 10.1007/1-4020-3312-5_42

[59] Barakat MN, El-Sammak H. In vitro mutagenesis, plant regeneration and characterization of mutants via RAPD analysis in Baby's breath *Gypsophila paniculata* L. Australian Journal of Crop Science. 2011;**5**:214-222

[60] Kaul A, Kumar S, Ghani M. In vitro mutagenesis and detection of variability among radiomutants of chrysanthemum using RAPD. Advances in Horticultural Science. 2011; **25**(2):106-111

[61] Barakat MN, Fattah RSA, Badr M, El-Torky MG. In vitro mutagenesis and identification of new variants via RAPD markers for improving *Chrysanthemum morifolium*. African Journal of Agricultural Research. 2010;**5**(8):748-757. DOI: 10.5897/AJAR09.679

[62] Kang EJ, Lee YM, Sung SY, Ha BK, Kim SH, Kim DS, et al. Analysis of the genetic relationship of gamma-irradiated in vitro mutants derived from standard-type chrysanthemum cv. Migok. Horticulture, Environment and Biotechnology. 2013;**54**(1):76-81. DOI: 10.1007/s13580-013-0124-9

[63] Wu JL, Wu C, Lei C, Baraoidan M, Bordeos A, Madamba MRS, et al. Chemical-and irradiation-induced mutants of indica rice IR64 for forward and reverse genetics. Plant Molecular Biology. 2005;**59**(1):85-97. DOI: 10.1007/s11103-004-5112-0

[64] Ahloowalia BS, Maluszynski M. Induced mutations–A new paradigm in plant breeding. Euphytica. 2001;**118**(2):167-173. DOI: 10.1023/A:1004162323428

[65] Wilde HD, Chen Y, Jiang P, Bhattacharya A. Targeted mutation breeding of horticultural plants. Emirates Journal of Food and Agriculture. 2012;**24**(1):31-41. DOI: 10.9755/ejfa.v24i1.10596

[66] Till BJ, Cooper J, Tai TH, Colowit P, Greene EA, Henikoff S, et al. Discovery of chemically induced mutations in rice by TILLING. BMC Plant Biology. 2007;**7**(1):19. DOI: 10.1186/1471-2229-7-19

[67] Cooper JL, Till BJ, Laport RG, Darlow MC, Kleffner JM, Jamai A, et al. TILLING to detect induced mutations in soybean. BMC Plant Biology. 2008;**8**(1):9. DOI: 10.1186/1471-2229-8-9

[68] Minoia S, Petrozza A, D'Onofrio O, Piron F, Mosca G, Sozio G, et al. A new mutant genetic resource for tomato crop improvement by TILLING technology. BMC Research Notes. 2010;**3**(1):69. DOI: 10.1186/1756-0500-3-69

[69] Caldwell DG, McCallum N, Shaw P, Muehlbauer GJ, Marshall DF, Waugh R. A structured mutant population for forward and reverse genetics in Barley (*Hordeum vulgare* L.). The Plant Journal. 2004;**40**(1):143-150. DOI: 10.1111/j.1365-313X.2004.02190.x

[70] Castillo AM, Cistue L, Valles MP, Sanz JM, Romagosa I, Molina-Cano JL. Efficient production of androgenic doubled-haploid mutants in barley by the application of sodium azide to anther and microspore cultures. Plant Cell Reports. 2001;**20**(2):105-111. DOI: 10.1007/s002990000289

[71] Kim DS, Lee IS, Jang CS, Kang SY, Seo YW. Characterization of the altered anthranilate synthase in 5-methyltryptophan-resistant rice mutants. Plant Cell Reports. 2005;**24**(6):357-365. DOI: 10.1007/s00299-005-0943-y

[72] Kim DS, Lee IS, Jang CS, Hyun DY, Seo YW, Lee YI. Selection of 5-methyltryptophan resistant rice mutants from irradiated calli derived from embryos. Euphytica. 2004;**135**(1): 9-19. DOI: 10.1023/B:EUPH.0000009509.78515.8e

[73] Hwang JE, Ahn JW, Kwon SJ, Kim JB, Kim SH, Kang SY, et al. Selection and molecular characterization of a high tocopherol accumulation rice mutant line induced by gamma irradiation. Molecular Biology Reports. 2014;**41**(11):7671-7681. DOI: 10.1007/s11033-014-3660-1

[74] Lal JP, Tomer AK. Genetic enhancement of lentil (*Lens culinaris* Medikus) for drought tolerance through induced mutations. In: Induced Plant Mutations in the Genomics Era. Rome, Italy: Food and Agriculture Organization of the United Nations; 2009. pp. 151-154

[75] Do KT, Dao MS, Hung PQ, Nguyen TC. Rice Mutation Improvement for Short Duration, High Yield and Tolerance to Adverse Conditions in Mekong Delta of Viet Nam; Plant mutation reports. Vol. 1, No. 1, 2006

[76] Tran DQ, Dao TTB, Nguyen HD, Lam QD, Bui HT, Nguyen VB, et al. Rice Mutation Breeding in Institute of Agricultural Genetics. International Atomic Energy Agency (IAEA). Viet Nam; 2006

[77] González MC, Pérez N, Cristo E, Rodríguez M, Borras O. Development of salinity-tolerant rice varieties using biotechnological and nuclear techniques. In: Induced Plant Mutations in the Genomics Era. Rome: Food and Agriculture Organization of the United Nations (FAO); 2009. pp. 138-140

Chapter 5

Water and Ecosystem Cycles Mediated by Plant Genetic Resources for Food and Agriculture

Masatoshi Funabashi

Additional information is available at the end of the chapter

http://dx.doi.org/10.5772/intechopen.79781

Abstract

Plant genetic resources for food and agriculture play essential roles in sustainable development and the conservation of global biodiversity. Especially, water cycle and related material circulation are deeply influenced by the loss of plant species diversity and external inputs through agricultural practices. This chapter overviews the water and ecosystem cycles mediated by the ecosystem functions of naturally occurring plant communities and discusses possibilities for the transformation of agriculture into sustainable modality with the primary importance on the recovery of water cycle. The transformation requires an intensive utilization of plant genetic resources in various ways compatible with a multi-scale integrated model of water and material cycles based on the processes of ecological succession and evolution. This foresight sheds light on the new importance and utility of plant genetic resources for food and agriculture, in the face of climate uncertainty and in repairing disrupted water and ecosystem cycles.

Keywords: water cycle, material cycles, biodiversity, ecosystem functions, plant genetic resources, ecological optimum, agriculture

1. Introduction

An abundant water cycle supports the ecosystem in the world we live in. Without water filling the earth's surface and flowing back to it, life would not have reached dry land and flourished. Water is the most important substrate for life. The underground water permeating the soil, the rivers flowing through the earth's surface, the lakes creating lush landscapes, and other components of the dynamic water cycle are all supported by the activities within the ecosystem. This chapter looks at the interactive relationship between the water cycle and the ecosystem.

From the total amount of water cycled across the entire earth, only 0.8% can be used for daily life, such as the underground water and the water in rivers and lakes. About 97% of the earth's water is seawater, and from the remaining 3% freshwater component, 2% is in the form of ice in glaciers in the polar regions. The water that we actually see in our daily life other than on sea is only a portion of the 0.01% water flowing on the ground, out of the 0.8% useable water [1].

This may be a very small number, but this 0.8% component of the water cycle supports the biodiversity on land. Our daily life, and industries such as manufacturing, agriculture, forestry, marine products, and livestock, is established from the various ecosystem services, including domestic water, derived from this component. Water flows in a form that we can use for daily life due to the power of the ecosystem, at the base of which are the plants.

Studying the impacts of agriculture by considering agricultural lands as artificially built ecosystems provides useful insights on the ecosystem and the water cycle. Conventional agricultural practices may be considered as a kind of disruption experiment on the water cycle that is mediated through the ecosystem. In addition, the kind of ecosystem that is created by agriculture is an important consideration in thinking about the future of the water cycle, which supports our daily life.

2. Soil functions, water retention, and water purification capacities afforded by the ecosystem

Let us look at how the ecosystem creates a water cycle that supplies water for our daily life. Plants, the primary producers of the ecosystem, play a central role in creating the water cycle for the miniscule 0.8% surface water component. Part of the rainwater supplied to the earth's surface by rainfall goes back to the atmosphere through evaporation and transpiration, and the rest repeatedly penetrates and flows out between the surface and the ground as it flows to the rivers. Within this process, plants—through photosynthesis—create the organic matter that becomes the source of energy for the entire terrestrial ecosystem. Other components of the ecosystem food chain, such as animals and fungi that degrade organic matter, are heterotrophs that depend on the organic matter produced by plants. Photosynthesis is the origin of all the processes on the earth's ecosystem. Thus, plants can be considered as the source of all organic matter on earth, including food and fossil fuel. Photosynthesis, by causing the formation of surface soil, serves as the driving force for the different cycling patterns of useable water, such as underground water, rivers, and lakes.

The electrical properties of surface soil provide clues for understanding the development of surface soil functions. Land, which was originally composed of rocks and minerals, is broken down into various sizes of gravel and eventually into fine clay through long years of erosion and weathering. In particular, since fine pieces of clay carry surface charges, they adsorb the ions needed for plant growth into the ground surface. In the same way that a piece of paper adheres to a sheet of plastic after producing static electricity by rubbing the surface of the plastic sheet, the ions contained in the rainwater are adsorbed by the clay and remain on the ground surface. The growth of plants on the ground and the resulting activities of microorganisms produce various organic substances, which further strengthen the electrical properties of the soil, thereby increasing its water retention capacity and the adsorptive power of microelements. Adsorption and retention of all sorts of substances by the earth's surface enhance the function of purifying the underground water permeating through the ground. As such, the electrical properties of the soil surface provide the key for the synergistic interaction between the production of organic matter and water retention and purification capacities of the soil.

Generally, the topsoil's water retention and purification functions increase proportionally with the diversity of vegetation. This is the reason for the preservation of virgin forests, which are composed of different tree species, as watersheds. Therefore, even under the same climatic conditions, the water cycle components available to living things vary greatly depending on the ecosystem formed on the ground surface layer. Even though the surface water component is only 0.8% of the water cycle for the entire earth, the circulation of this component has a significant impact on the habitat of terrestrial organisms. Imagine what would happen if there was no vegetation on earth, and there was no formation of soil due to photosynthesis. There would be no underground water, rivers, or lakes that would provide a ready supply of water for life. Rainwater would flow straight into the ocean, land would simply release rocks into the sea by erosion, and this world would be nothing but either raging streams of water or dried valleys.

3. Merits and demerits of agriculture: decline of usable water cycle components

We need to reevaluate agriculture in consideration of the general role of the ecosystem in soil formation. Usual agricultural practices involve the tillage of land to enhance water retention. This results in a sure but short-term increase of water retention in tilled agricultural lands, sometimes to a level similar to that of forest areas [2]. However, unless tillage is regularly and continuously done, the water retention capacity will not be sustained; that is, without continuous tillage, a hard crust will be formed on the surface to repel rainfall, and the capacity to retain water will eventually be lost. Also, since tillage disturbs the soil ecosystem, it will also eventually destroy the capacity of the topsoil to purify water. In reality, agricultural fertilizers are the biggest pollutants being released to rivers. There are reports of cases wherein, depending on crop species, more than 90% of the nitrogen, phosphate, and potassium—the major components of fertilizers—are not absorbed but become runoff, in agricultural practices that entail tillage and use of fertilizers. This percentage is not for only some farms, but represents the national average in Japan [3].

The emphasis on temporary water retention in conventional agricultural practices clearly does not take into consideration the environmental impact on the ground surface water cycle. Addressing this issue means having to deal with the complexity of the water cycle. First of all, underground water contamination caused by agriculture, unlike industrial and domestic effluent, does not have exact discharge locations, making it difficult to identify testing and sampling points. Also, due to the complex dynamics of underground water penetration, identifying clear causal relationships is not easy, preventing the analysis of the interplay of different farms and timings of fertilization that lead to contamination. If the contamination has accumulated along the river basin, the effects to distant areas, including the marine ecosystem, must also be considered. The complexity of the water cycle widens the extent of the impact, thereby obscuring the location of responsibility for contamination.

The most typical example I witnessed in the field is the runoff of red clay in sugarcane fields in Ishigaki Island. With subsidy from the government, farmers grow sugarcane by tilling land in summer when rainfall is heaviest on 1700 of the 22,900 hectares of the total island area [4]. From an airplane, you can clearly see the red clay flowing in all direction throughout the island and the coral reefs turning reddish brown in color during rain. Fertilizer-containing red clay causes significant changes to the environment of organisms living in the coral reefs, which are said to be responsible for 80% of the biodiversity of marine ecosystems. Recently, there has been an abnormal proliferation of crown-of-thorns starfish around Ishigaki Island, causing damage to the corals. Remains of dead crown-of-thorns starfish are washed up on the beaches, making some areas dangerous to walk barefoot because of their poisonous spines.

Long-time residents of Ishigaki Island claim that they have not seen such occurrences in the past few decades. Although crown-of-thorns starfish is important in creating the diversity of corals, eutrophication due to fertilizer runoff has caused its abnormal proliferation [5].

Thus, although it is clear that agriculture is affecting the environment, it is difficult to make comparisons to identify exact causal relationships and determine the extent of the effects. The selection of factors that must be quantified in order to understand the phenomenon also depends on the purpose and the scale of investigation. Understanding the effect of the coral reef on the ecosystem would require investigating an extensive range of factors that include the fluctuations in marine biodiversity. Likewise, finding correlations between climatic and agricultural factors and isolating individual effects would entail very complicated processes. Therefore, prior to quantifying the complex dynamics of the problem, it would be more effective to qualitatively identify upstream factors and remove them from the targeted systems.

Many years ago, environmental contamination from pesticides, rather than from fertilizers, was the more urgent concern. Pesticides, which have direct toxic effects, more easily became subject to environmental and ethical discussions. At present, most of the pesticides used are highly degradable and do not leave residues in the environment. Even pesticides with low toxicity, however, when used in the long term or in combination, lead to indirect as well as direct effects on the ecosystem—effects that cannot be determined in advance. Likewise, even though fertilizers are in themselves not toxic, they diminish useable water resources once they enter the water cycle, leading to reduction in biodiversity of the water ecosystem, loss in income from fisheries, and damage to the water-related living environment. Also, when using organic fertilizers from livestock farms, there is a need to consider the risk of releasing antibiotics and other chemicals used for animals into the water ecosystem.

Therefore, before the contamination spreads throughout the complex components of the water cycle, the basic surface soil functions must be preserved, and fundamental measures must be implemented in agriculture to prevent the creation of contaminants in the first place. An example of these measures is the incorporation of cover cropping (planting of grasses and legumes in between cropping to prevent soil erosion, enhance the landscape, or suppress the growth of weeds) as part of conventional agriculture practices. Also, since chemical fertilizers and pesticides rely on petroleum resources and rare metals, they cannot be supplied sustainably. Therefore, as long as we do not make use of the natural water retention and purification capacities of the water cycle, which are underpinned by healthy ecosystem functions, the cost for investing on artificial measures would be too high.

The use of genetically modified crops, which have continued to improve in recent years, is gaining wide attention as a means to increase yield while decreasing inputs and environmental burden. This is an effort to shift from the control of environmental factors through the input of material resources to the manipulation of genetic functions. The basic framework of agricultural methods, however, still entails the destruction of soil functions through tillage. In other words, the priority lies in optimizing agricultural productivity under the conventional framework, without consideration of the water cycle functions of the earth's surface afforded by the ecosystem. As such, there are at most only around ten types of environmentally related genes incorporated in genetic modification of crops. This is in stark contrast to the innumerable number of genes that are related to the wide range of ecosystem functions supporting the water cycle and that are expressed by all organisms involved in the formation of surface soil.

It is therefore more important to figure out how to allocate vegetation that supports the core of the water cycle, which involves a numerous number of genes, rather than create a crop that incorporates around ten new genes, in thinking about agriculture that contributes rather than undermines the water cycle. The diversity of the countless genes expressed by plants, animals, and microorganisms in response to the environment is the foundation that supports the water cycle at the genetic level. The interspecific transfer of genes to the surrounding ecosystem through genetically modified crops, however, can have a negative effect on this diversity, and its actual risks are still unknown. In the same way that vigorously introduced species sometimes impair the diversity of indigenous organisms, there is also a risk of diminishing the genetic diversity of endemic species through hybridization with genetically modified crops endowed with dominant functions.

We need to give careful thought on whether it is worth the risk of adversely affecting the diversity of the immensely abundant genetic resources underpinning the water cycle for the sake of optimizing single-crop farming, which is based on the destruction of soil functions. Moreover, even if we can provide proof of whether the transfer of genes from genetically modified crops to the surrounding ecosystem has occurred in the past or not, it is in principle impossible to guarantee that it would not happen in the future.

4. Embankments and flood control

River embankments are an important consideration in thinking about the water cycle of rivers, which are important in agriculture. Rivers overflow by nature, and the riverbanks naturally formed from the flooding, as well as the fertile floodplains around the riverbanks, is in fact ecological structures that naturally create the river's water cycle. The vegetation on the floodplains is what sustains the underground water and functions to store the excess water during flooding, other than being an important source of biodiversity. The fertile floodplains near the rivers are suitable for agriculture and building cities. Due to the resulting advanced economic growth along the rivers, concrete embankments were built along the banks of rivers all throughout Japan to develop the alluvial plains along them. Meanwhile, the growth of the cedars and cypresses that were planted on the mountains in different areas in Japan after World War II has led to the decline of forest floor vegetation, making the mountains more prone to landslides. This has in turn led to the overbuilding of concrete embankments even for the small upstream rivers of mountains.

Cemented riverbanks at a glance seem to protect us from flooding of rivers, but they in fact undermine the diverse water cycle and ecosystem functions of rivers. The water retention capacity and the many other benefits brought about by rivers to the surrounding ecosystem were lost as a result of cementing the passage of water to contain the rivers. Without embankments, the water in rivers freely flows to and from the nearby underground water sources while being filtered through the diverse soil environment. The purification process is compatible with the principle of septic tanks used for the domestic effluent such as from toilets, etc., which are based on physical filtration and adsorption and degradation of organic matter by both aerobic and anaerobic microorganisms. The more diverse the soil environment is, the higher is its capacity to purify water passing through it. The river basin water is purified as it passes through diverse soil types of different physical environments and microbial flora while at the same time enhancing their water-purifying functions.

Cemented riverbanks intercept the free flow of underground water and undermine the inherent purifying function of riverbanks. The Miya River flowing through Ise, Mie Prefecture, is one of the few Class 1 rivers in Japan. According to people living near the river's estuary, before the concrete riverbanks were built, there was no sewage system, there was a large population of residents, and there were some kitchen scraps and garbage floating around. The water, however, was clear, and people were able to dive into the river and catch fish. Presently, however, the population of residents near the estuary has declined, and while a sewage system has been put in place, sludge has built up on the bottom of the water channel and has created a stench. Since disposal of domestic and industrial effluent is restricted, the water quality problem is believed to be caused by fertilizer runoff from the upstream farms and by the loss of purifying capacity of the cemented riverbanks. Also, the decline in the population of eel, which has recently been classified as an endangered species, is believed to be caused, other than by overfishing, by the loss of their habitat due to the building of concrete riverbanks across Japan [6].

Thus, impediments to the natural water cycle result in various trade-offs in ecosystem functions, leading to risks of forfeiting the aggregate benefits afforded by rivers. Since rivers change their courses within the floodplain in response to the water cycle, restricting the flow of rivers at the convenience of human society would only be good for several decades. When we consider the movement of rivers over a hundred-year span, it is possible that the cost of floods that cannot be prevented by the concrete riverbanks and the ecosystem functions lost by building them would exceed the benefits of building concrete embankments. Even the cities and farmlands damaged by the tsunami during the Great East Japan Earthquake were alluvial plains close to the sea level (some were reclaimed areas built lower than the sea level), which basically serve as buffers of the effects of changes in the water cycle.

The Netherlands, Germany, Austria, and other countries have implemented flood control measures that take ecosystem functions into consideration through renaturalization of rivers by removing the embankments and restoring the floodplains. Conventional flood control by building concrete riverbanks has led to the worsening of environmental deterioration and loss of biological resources. Through citizen's movements and policy decisions, consensus is building toward solving these problems by allowing the water cycle of rivers to take its natural course [7].

Large dams built across the USA are approaching the end of their lifespans, and it has been pointed out that they have in fact not fully performed their intended functions in generating power, irrigation, and flood control. Rather, they have adversely affected water quality and the renewal of resources, reduced the underground water in the river basins, and restricted the movement and habitat of wildlife. Because of this, around 850 dams have been demolished in the last 20 years [8]. The dams built around the world are able to hold up to 15% of the total water flowing in all the rivers on earth—an indication of the magnitude of the effect of artificial water reservoirs on the water cycle on the surface of the earth [9].

The renaturalization of rivers is based on the recognition of the environmental deterioration caused by man's efforts to control rivers by artificial means. The concept of renaturalization also includes the extreme approach of completely letting nature take its course by disengaging from all human activity, such as agriculture, in the floodplains. Many aspects of the relationship between ecosystem functions and the water cycle in floodplains still remain unexplained. As the renaturalization of rivers continues to progress, it is therefore necessary to reassess its benefits and shift to a new form of industrial activities that maximize those benefits.

5. Agriculture that promotes the ecosystem functions related to the water cycle

Thus far, we have looked at examples of the major effects of both the natural cycling and artificial cycling of water on the ecosystem. The artificial water cycle created by dams and embankments was originally intended to enhance the production of drinking water, energy, and food needed by humans. Since they were built without regard for the effects on the ecosystem and on the natural water cycle, however, they have led to various ills as well. The renaturalization of rivers is a movement to return to the original course of nature upon the realization that the adverse impacts of development are far larger than its benefits. But, is there an example in which the artificial modification of the water cycle has positively enhanced both agricultural productivity and biodiversity?

There is a vast expanse of paddy fields around San Francisco in the USA. The area used to be dry like a desert, but a dam was built to collect water from melted snow from the Sierra Nevada mountains and irrigate the fields, enabling regulation of the water levels to the millimeter level and the cultivation of rice. Applying different concentrations of fertilizers depending on the previous year's harvest enables minimizing fertilizer runoff, and the percolation is blocked by the underlying bedrock. Even though rice paddies are man-made marshlands, they provide a habitat for ducks and various waterfowls and marsh animals. Thus, along with harvesting and other agricultural practices, local volunteers also conduct wildlife conservation activities to protect the young birds and other animals. It is therefore possible to enhance an aspect of biodiversity along with agricultural productivity, even within an artificially created water cycle. Globally Important Agricultural Heritage Systems (GIAHS) represent examples of agriculture that balances productivity and conservation in different countries around the world, wherein moderate disturbance by humans enhances the biodiversity of the environment, such as Satoyama farming (traditional farming in Japan at mountain skirts near the villages) [10, 11].

Would it possible, therefore, to more widely practice agriculture that is based on the relationship between biodiversity and the cycling of water and other resources within the ecosystem? Thus far, agriculture focused too much on continuous production of a particular type of crop, which required a different supply pathway from the natural system for the cycling of materials in the ecosystem, and could not be produced sustainably without the input of external resources. This is very similar to the destruction of floodplains of rivers contained by concrete embankments. The damage from continuous cropping, which is a normal agricultural practice, can be avoided by allowing the succession of vegetation in a natural ecosystem, which is premised on the mixture of a variety of species. In the same way that the natural ecosystem functions of rivers can be restored by enabling flooding to take place, would it be possible to restore the autonomous functions of ecological succession in agriculture?

Let us try to think of an agricultural production system based on the process of topsoil formation, the starting point of which is photosynthesis. This production system should effectively utilize the water cycle, which is nurtured by the process of topsoil formation. Going back to how the cycling of materials has evolved, we see that each material cycles between the various layers comprising the ecosystem. At the beginning, after exposure to rainfall and sunlight, the rocks and minerals were eroded and dissolved into soil solution, where microorganisms started to deposit organic matter, leading to the growth of plants on the first layer of soil formed. Eventually, animals arrived to take up a higher position in the food chain (**Figure 1**). The current composition of the atmosphere is a result of the total effects of the earth's ecosystem, which has been modified by the evolution of living organisms. Being formed from the lower layers beneath them, the higher layers are more complex, so that the layers can be arranged vertically based on their complexity. We will refer to this arrangement as "axial hierarchy." **Figure 2** illustrates how the cycling of materials between each evolution layer takes place. The succession of the ecosystem and the soil forms a network that is intricately entwined with the cycle of materials. By assigning a qualitative complexity score to the soil and ecosystem succession based on the stage of emergence and evolution, and averaging the scores of the layers related to each material cycle, we can arrange them based on the characteristics of emergence and evolution. The resulting order shows a close correlation between the evolution of the ecosystem and the cycling of materials in accordance with the acquired characteristics: In **Figure 2** right, the lowest complexity score of material cycle that is

http://dx.doi.org/10.5772/intechopen.79781

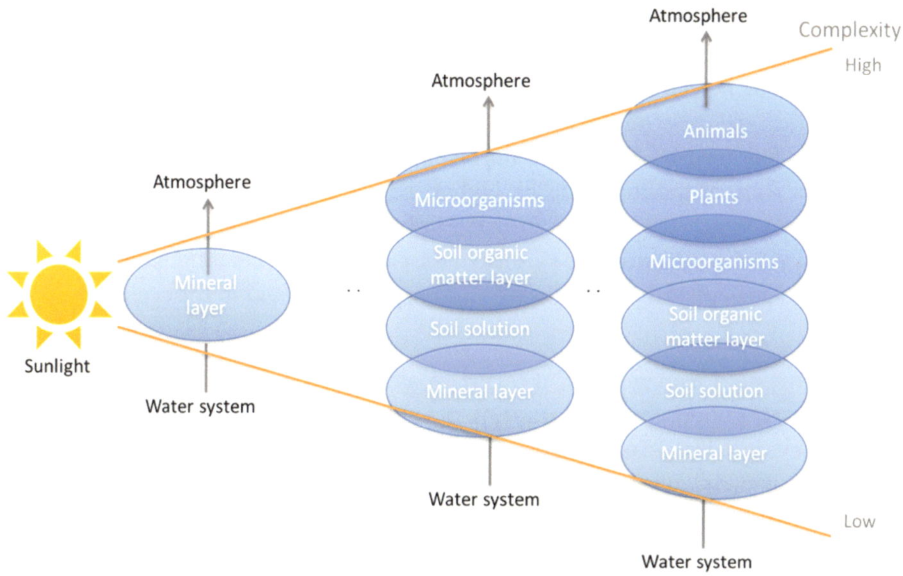

Figure 1. Axial hierarchy of emergence and evolution of land ecosystems.

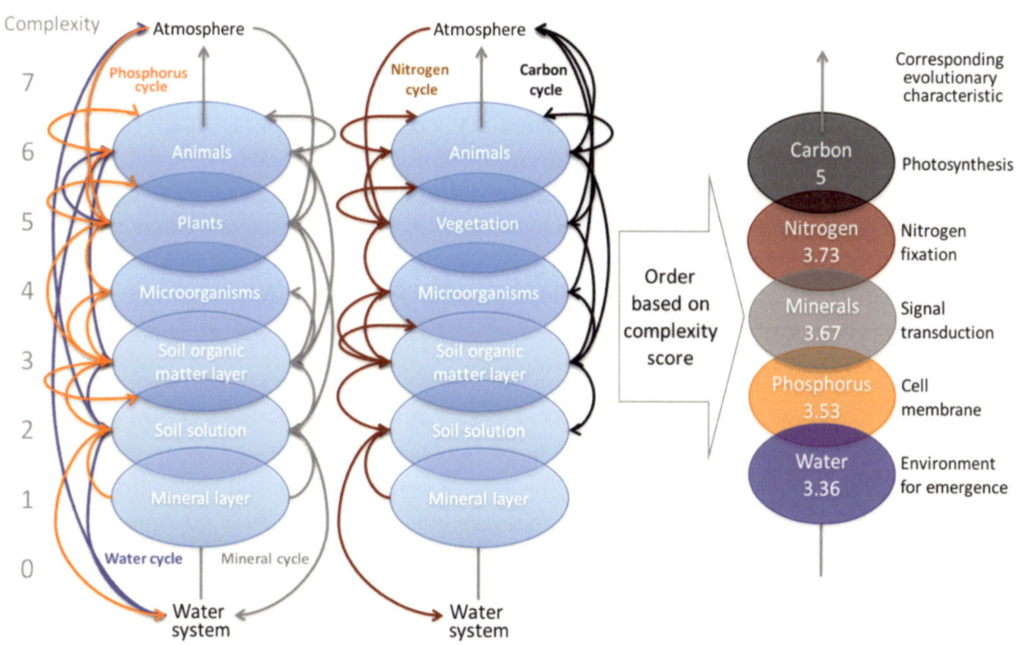

Figure 2. Axial hierarchy of material cycles of land ecosystems. Prepared based on the Figure 18-2 in Ref. [21] by adding the animal layer and leguminous nitrogen fixation. Water movement between animals and transpiration of water from animals were excluded from the relative amount of water.

most involved in the initial process of land ecosystem evolution is the cycle of water, which provides the environment for the emergence of life. Then, on the upper layer followed by the cycle of phosphorus, which is responsible for cell membrane formation, followed by the cycle of minerals, which mediate the signal transduction systems needed for the independent activities of the cell, followed by fixation of nitrogen, by which *Archaebacteria* carry out nutrient transformation. And, at the top with the highest complexity score of land ecosystem, is the fixation of carbon, which is carried out through photosynthesis, which is in turn responsible for the vegetation covering the entire surface of the earth and for creating the topsoil. Thus, the cycling of materials depicts the evolutionary history entwined within the ecological succession process.

In terms of ecosystem management, if a certain element in the hierarchy is lacking, then a problem occurs. Traditional agriculture tries to solve the problem by supplying the lacking element. However, if the level supplied by the natural cycle is lower than the artificially supplied level, then the system cannot be sustained unless the input is continually made. Conversely, to realize a sustainable culture systems, we need to think about how to change the ecosystem to enable a natural remediation of the material cycle without supplying the lacking element. By fixing the ecosystem layer where the cycle of the lacking element is mostly occurring, it is possible to enhance the cycle of that element within the natural cycle.

For example, to enhance the water cycle, you can plant vegetation with high biomass on fine-textured soil to increase the accumulation of organic substances on the topsoil and improve water retention. This increases the volume of water retained by the topsoil for the same amount of rainfall. Also, since the purifying function of the topsoil is dependent on the thickness of the topsoil formed by the vegetation, the capacity to decompose organic substances in the topsoil also increases. Likewise, to increase the amount of phosphorus, minerals, and other microelements, you can plant tree species and vegetation that attract insects and small animals carrying those elements to the field and collect them from surrounding ecosystems. These microelements, other than being supplied through weathering of minerals and through rainfall, are dispersed and retained on land through the activity of fish-eating birds and other animals that collect them from sediments originating from the oceans. Meanwhile, nitrogen, carbon, and other important components of organic matter are accumulated through photosynthetic activities by plants and through the activities of symbiotic microorganisms; therefore, succession of vegetation can be allowed to proceed until the necessary soil formation level is reached.

These measures, however, cannot be expected to immediately satisfy the recommended rates of fertilizer application for single cropping of particular varieties selected by modern agriculture. In contrast, they can enhance the biodiversity of microelements and phytochemicals that are important for exhibiting the ecosystem's functions. It is therefore possible to sustain productivity based on the ecological optimum under mixed growth conditions, which is the basis for primary productivity in natural ecosystems [12].

In order to increase productivity in a highly diverse plant community, there is a need to find the right set of useful plants that grow under the niche of that community rather than practice continuous cultivation of a specific crop. This is the same as improving the cycling of materials by changing their relationships rather than by input of lacking elements. It is possible to balance productivity with the natural ability of the ecosystem to adapt to environmental changes by designing vegetation to enable productivity in a diverse community, in accordance with the cycle of materials and the environment established in the field. Agricultural crops are basically introduced species, wherein despite having more than 30,000 species of agriculturally useful plants, there are only around 120 species actually being cultivated for agriculture. About 90% of all food is derived from only 30 species of plants [13]. Enhancing the water cycle and other ecosystem functions based on biodiversity requires the development and cultivation of underutilized or neglected plant resources in accordance with the ecological succession stage. Synecological farming, or synecoculture, is such an approach to agriculture that emphasizes the management of ecological relationships.

In synecoculture, an extremely wide variety of useful plants are densely cultivated together based on the ecological optimum. This results in a condition wherein the water retention capacity and permeability of the topsoil enhance the water-buffering capacity of the soil while supporting the growth of aboveground vegetation and at the same time enhancing the biodiversity and activity of soil microorganisms [14]. Field experiments conducted in Burkina Faso, sub-Saharan Africa, showed that synecoculture was more efficient in the consumption of water relative to productivity compared to other cultivation methods [15]. These examples indicate that the intensive introduction of diverse genetic resources from useful plants into an agriculture based on the ecological optimum is very effective in improving ecosystem functions, such as the cycle of water and materials.

6. Target areas for implementing agriculture that is adaptive to fluctuations in the water cycle

In particular, which places would benefit from an agriculture that emphasizes uninhibited ecological dynamics and the conservation of the water cycle? Governments, NPOs, and scientists from more than 110 countries have submitted an international report stating that agriculture based on large-scale monocropping is not sustainable from the standpoint of environmental burden and fair distribution of food [16]. Also, those who are at higher social risk against the effects of climatic changes are not the developed nations with advanced large-scale agricultural systems, but the small and developing countries in the tropical and subtropical regions [17]. In particular, the increasing expanse of flood-stricken areas in Southeast Asia is at a high risk of being unable to cope with dramatic changes in the water cycle if conventional agriculture is continued, pointing to the urgent need for developing agriculture that leverages the inherent water-buffering capacity of the ecosystem. In China, 200 million small-scale farms are pursuing modernization with support from the government. There is a need, however, to implement measures to improve food production based on the ecological optimum rather than through conventional agricultural practices, both from the standpoint of environmental burden and the available materials and resources in the future [18]. In India, where the Green Revolution has since steadily increased food production and enabled overcoming hunger, restoring the biodiversity and ecosystem services (particularly pollination and purification of water and air) that were lost as a result of the agricultural activities must be addressed [14]. Moreover, African countries undergoing rapid development in recent years must develop self-sustaining agriculture practices based on the characteristics of the ecosystem in each region, in order to combat desertification in arid areas, correct the disparity in wealth, and stabilize their society [15, 19]. To arrest the deforestation of the few remaining tropical rainforests in Indonesia, the Amazon in South America, the Congo Basin, and other susceptible areas, we need to nurture the capacity of local communities to use forest resources sustainably based on the ecological optimum of useful plant resources. We also need to create regulations that favor local economic development and involve stakeholders from a wide range of sectors. These regions will benefit from the rapidly increasing uptake of mobile terminals and other IT devices through the development of databases and tools for understanding relationships with the water cycle and biodiversity primarily pertaining to useful plant resources. These databases and tools will serve as effective infrastructures for underpinning next-generation ecosystem management and food productivity [20].

7. Conclusion

This chapter reviewed the relationship between biodiversity and ecosystem cycles in plant communities with a particular focus on water cycle. A wider introduction of plant genetic resources for food and agriculture into the establishment of novel agricultural practices that make use of ecologically optimum formation of mixed communities is needed to overcome sustainability burdens, cope with the climate change, and recover globally disrupted water and material cycles. Guidelines for the resolution through ecosystem management are explained along with the "axial hierarchy," as a structure traversing ecosystem cycles closely related to the emergence and evolution of land ecosystems.

Acknowledgements

This research is supported by Sony Computer Science Laboratories Inc. and the Center of Innovation Program "Global Aqua Innovation Center for Improving Living Standard and Water-sustainability" from the Japan Science and Technology Agency, JST.

Conflict of interest

The author declares no conflict of interest.

Author details

Masatoshi Funabashi[1,2*]

*Address all correspondence to: masa_funabashi@csl.sony.co.jp

1 Sony Computer Science Laboratories, Inc., Tokyo, Japan

2 Japan Agency for Marine-Earth Science and Technology, Yokosuka, Japan

References

[1] Shiklomanov IA. Assessment of Water Resources and Water Availability in the World. Geneva, Switzerland; World Meteorological Organization (WMO); 1996

[2] Fujimura K. Stormwater Runoff Analysis of Urban Basins Using a Rainfall Simulator to Measure Infiltration, and Its Application [Thesis], Tokyo Metropolitan University, Tokyo, Japan; 1999. [in Japanese]

[3] Nishio M. Agriculture and Environmental Pollution (Nougyou to kankyou osen) – Soil Environment Policies and Technologies in Japan and the World. Tokyo, Japan: Rural Culture Association Japan; 2005. p. 65-73. [in Japanese]

[4] Support for cultivation of millet stock [Internet]. 2010. Available from: http://www.strata.jp/sangokikin/report/101008report.pdf [Accessed: 2018-05-25] [in Japanese]

[5] Bell PRF. Eutrophication and coral reefs—some examples in the Great Barrier Reef lagoon. Water Research. 1992;**26**:553-568. DOI: 10.1016/0043-1354(92)90228-V

[6] Itakura H, Kimura S.Reduction in eel population in areas enclosed by embankments. Nature Conservation. Tokyo, Japan: The Nature Conservation Society of Japan; 2014;**540**: 24 [in Japanese]

[7] Hayano H. Rivers and Europe: River Renaturalization Concept. Tokyo, Japan: Tsukiji Shokan Publishing; 2003. [in Japanese]

[8] Rummel T (produce), Knight B, Stoecker M, Chouinard Y, Calhoun B. Damnation. Ventura, California, United States: Patagonia Books; 2014. DVD

[9] Black M, King J. The Atlas of Water: Mapping the World's Most Critical Resources. [Translated into Japanese by Oki T and Oki A]. 2nd Edition. Maruzen, Tokyo, Japan; 2010

[10] GIAHS [Internet]. Available from: http://www.fao.org/giahs/en/ [Accessed: 2018-05-25]

[11] European Environmental Agency. High nature value farmland Characteristics, trends and policy challenges. EEA report No. 1/2004; 2004

[12] Putman RJ, Wratten SD. Principles of Ecology. Oakland, California, United States: University of California Press; 1984

[13] Yong RN, Mulligan CN, Fukue M. Geoenvironmental Sustainability. Boca Raton, Florida, United States: CRC Press; 2006. p. 168

[14] Funabashi M. Synecological Farming for Mainstreaming Biodiversity in Smallholding Farms and Foods: Implication for Agriculture in India. Indian Journal of Plant Genetic Resources. 2017;**30**(2):99-114. DOI: 10.5958/0976-1926.2017.00016.X

[15] Tindano A, Funabashi M (eds). Proceedings of the 2nd African forum on Synecoculture. Research and Education material of UniTwin UNESCO Complex Systems Digital Campus, e-laboratory: Open Systems Exploration for Ecosystems Leveraging, No. 7; 2017

[16] International assessment of agricultural knowledge, science and technology for development (IAASTD). Synthesis report, with executive summary: A synthesis of the global and sub-global IAASTD reports. IAASTD; 2009. Available from: http://www.agassessment-watch.org/report/Synthesis%20Report%20(English).pdf

[17] Petherick A. A note of caution. Nature Climate Change. 2012;**2**:144-145. DOI: 10.1038/nclimate1423

[18] Zhang F, Chen X, Vitousek P. Chinese agriculture: An experiment for the world. Nature. 2013;**497**:33-35. DOI: 10.1038/497033a

[19] Gaiser T, Judex M, Igué AM, Paeth H, Hiepe C. Future productivity of fallow systems in Sub-Saharan Africa: Is the effect of demographic pressure and fallow reduction more significant than climate change? Agricultural and Forest Meteorology. 2011;**151**:1120-1130. DOI: 10.1016/j.agrformet.2011.03.015

[20] Funabashi M. Open Systems Exploration: An Example with Ecosystems Management. In: First Complex Systems Digital Campus World E-Conference 2015. New York, United States: Springer International Publishing; 2017. p. 223-243. DOI: 10.1007/978-3-319-45901-1

[21] The Japanese Society of Forest Environment (eds). Balance in the Forest: Interaction of Plants and Soil (Mori no baransu). Hiratsuka, Japan: Tokai University Press; 2012. p. 209 [in Japanese]

Chapter 6

Genetic Variability of Mountain Pine (*Pinus hartwegii* Lindl) in the Protection of Flora and Fauna Area Nevado de Toluca

Rosa Laura Heredia-Bobadilla,
Guadalupe Gutiérrez-González,
Amaury-M. Arzate-Fernández and
Sergio Franco-Maass

Additional information is available at the end of the chapter

http://dx.doi.org/10.5772/intechopen.79860

Abstract

Mountain pine (*Pinus hartwegii* Lindl) is one of the most abundant conifers in the Protection of Flora and Fauna Area Nevado de Toluca in central Mexico; this natural protected area is threatened by urbanization; this has been manifested in forest health; there has been an increase in forest parasites like bark beetles and dwarf mistletoes, making necessary improve forest management and conservation, hence our objective was to study the genetic diversity of mountain pine under the attack of parasites and to generate information that could be used to improve strategies of conservation of these forests. We classified sampled trees into four categories according to the type of parasite present in a tree (bark beetle: BB; dwarf mistletoe: DM; bark beetle and dwarf mistletoe: BM and non-attacked trees or healthy trees: HT). Genetic diversity was low in comparison with other pine species, but we observed an interesting issue: trees attacked by bark beetle and dwarf mistletoe had higher levels of heterozygosis: $He_{nc} = 0.1924$ and $He_{nc} = 0.1993$, respectively. These results suggest that trees with bark beetle and dwarf mistletoes may have higher genetic variability and are a highly valuable genetic resource for mountain pine.

Keywords: mountain pine, genetics, bark beetle, dwarf mistletoe, conservation

IntechOpen

1. Introduction

Forests of the Trans-Mexican Volcanic Belt (TMVB) physiographic region are among the most threatened areas in Mexico. The TMVB region encloses a large number of natural protected areas (NPA), the Protection of Flora and Fauna Area Nevado de Toluca (PFFANT) is one of them. The closeness of this NPA to large cities entails environmental pressures that includes the presence of human settlements, illegal logging, introduction of exotic species and the growing incidence of forest parasites like bark beetles and dwarf mistletoes [1–3].

PFFANT is formed by conifer forests, with the genera *Pinus* and *Abies* being the most representative of this zone [4, 5]. *Pinus hartwegii* Lindl. is forming large forest stands and is the pine species which grows at the highest altitudes in the PFFANT [6], unfortunately is affected by dwarf mistletoes (*Arceuthobium* spp.) and bark beetles (*Dendroctonus* spp.). The incidence of these parasites is growing, probably because of climate change and deforestation [1, 7–9].

México ranks fourth in terms of deforestation, with approximately 670,000 ha/year causing losses in genetic diversity and changes in locally adapted populations for example giving place to an increase of parasites populations [10–13]. Results of forest health diagnoses have suggested that bark beetles are a group of parasites that have affected large extension of forests in Mexico (40.5%), followed by parasitic plants (38.7%), both of them leaving negative consequences in the forest, like high mortality rates of trees affected [11, 14, 15].

Bark beetles grow under the cortex and induce weakening of the tree, the construction of galleries and the inoculation of a staining fungi which is carried by female beetle results death [14]. In the PFFANT, *Dendroctonus adjunctus* is affecting *P. hartwegii* [6, 14]. Dwarf mistletoes are obligate heterotrophic plants that acquire all their water and nutrients from their host and can significantly inhibit its growth causing permanent deformation of the stem and crown. This parasite weakens trees in such a way that they become more susceptible to attack by insects, particularly bark beetles. In the PFFANT, *P. hartwegii* is the host of *Arceuthobium vaginatum* and *A. globosum* [6, 8, 15]. Bark beetles and dwarf mistletoes epidemics can lead to shifts in forest, forest successional trajectories and susceptibility to future disturbances [16–18].

Forest trees are key drivers of terrestrial biodiversity because they function as a carbon sink, preserve the water quality and regulate climate [19, 20]; genetic variability studies are crucial to understand the basic biology of these organisms and to obtain insights on evolution, disease resistance and conservation genetics [21–23]. In conifer species for example, gene flow is mediated by three types of genomes with contrasting inheritance: nuclear (biparental), mitochondrial (maternal) and chloroplast (paternal) this particularity opens avenues to the study of conifer DNA polymorphism, the study of genetic variability with these three types of markers allow making inferences on the distribution of genetic resources and habitat connectivity [24].

In this study, nuclear DNA (ncDNA), mitochondrial DNA (mtDNA) and chloroplast DNA (cpDNA) were used to assess the genetic variability and population structure of *P. hartwegii* populations affected by bark beetles and dwarf mistletoes, expecting that genetic variability will be low in the parasitized populations. We want to contribute to the conservation of mountain pine populations of the PFFANT generating information which helps in the identification of populations genetically valuable.

2. Materials and methods

2.1. Study zone

Sampling was carried out in the Protection of Flora and Fauna Area Nevado de Toluca, which forms part of the Trans-Mexican Volcanic Belt physiographic region. The geographical coordinates of the study zone are 18° 51′ 31″ and 19° 19′ 03″ N and 99° 38′ 54″ and 100° 09′ 58″ W; it is a priority region for conservation due to its diversity of ecosystems in which pine-fir forests and high mountain prairie dominate [1, 25].

2.2. Plant material

We sampled a total of 180 individuals of *P. hartwegii*. A distance of 50 m was between each tree sampled. Samples were classified into four groups (categories): trees with signs of attack by bark beetles (BB), trees with signs of attack by dwarf mistletoe (DM), trees with signs of attack of both parasites (BM), and trees with no signals of any parasite, which were considered as healthy trees (HT). Each sample consisted of young needle tissue. Immediately after collecting the needles, they were placed on ice for transport to a laboratory where they were maintained at −70°C.

2.3. DNA isolation

Cetyltrimethylammonium bromide (CTAB) method was implemented with a few modifications [26]. Needle tissue was ground to a fine powder with a chilled mortar and pestle, making two washes, the first with 100% ethanol (v/v) and the second with 75% ethanol (v/v). DNA was re-suspended in 70 μL TE and stored at −20°C until it was used.

2.4. PCR amplification reactions

PCRs were performed in a reaction volume of 10 μL, containing: ammonium buffer 15 mM, $MgCl_2$ 25 mM, dNTP mix 10 mM, primer 20 μM, DNA 10 ng/μL and 0.5 U of Taq polymerase (Sigma). To amplify nuclear (ncDNA), we used anchored microsatellites (ASSR) which proved to be genetically stables and heritable: ASSR-15 and ASSR-29 [27] and one operon: UBC 254 [28], one cytochrome oxidase sequence: cox3in, was used to amplify mitochondrial (mtDNA) [30] and one highly polymorphic microsatellite: 10FF/RR, to amplify chloroplast (cpDNA) [31] (**Table 1**).

Amplifications were performed in a Master Cycler Gradient (Eppendorf) Thermal Cycler in all 40 cycles. For ncDNA primers: 1 min of denaturation at 94°C, 1 min of annealing at 48°C and 1 min of extension at 72°C. For mtDNA primers: 1 min of denaturation at 94°C, 1 min of annealing at 48–55°C and 1 min of extension at 72°C. For cpDNA: 1 min of denaturation at

Primer name	Marker type	Sequence (5′ → 3′)
UBC-254	Nuclear operon [28]	5′-CGCCCCCATT-3′
ASSR-15	Nuclear anchored microsatellite [29]	5′-(CT)7GCA-3′
ASSR-29	Nuclear anchored microsatellite [29]	5′-(CT)7GTA-3′
ASSR-20	Nuclear anchored microsatellite [29]	5′-(CT)7ATG-3′
COX3in	Mitochondrial cytochrome oxidase III gene sequence [30]	5′-GTA GAT CCA AGT CCA TGG CCT-3′ 5′-GCA GCT GCT TCA AAG CC-3′
10 FF/RR	Chloroplast microsatellite [31]	5′-CAGAAGCCCAAGCTTATGGC, 5′-CGGATTGATCCTAACCATAC

Table 1. Description of markers used for the study of genetic variability of *Pinus hartwegii*.

94°C, 1 min of annealing at 58–60°C and 1 min of extension at 72°C; all samples with all primers were given a 7 min of pre-amplification denaturation at 94°C and a 5 min of post-amplification at 72°C.

2.5. Electrophoresis in agarose gel

The amplification products were separated by electrophoresis in agarose gel (1.5%) at constant voltage (100 V and 90 mA). Gels were visualized by UV transilluminator (UVP) with ethidium bromide (10 mg/mL).

2.6. Scoring of bands and data analysis

DNA patterns were inferred according to dominant nature of markers used, so each amplified product was scored for all genotypes for its presence or absence, we made a binary matrix in which band presence was assigned a value of one (1) and the absence of a band a value of zero (0). Co-migrating bands were assumed as the same locus and the same band when scoring.

We used PopGene 32 [32], Genealex 6.5 [33] and TFPGA [34] to obtain genetic diversity parameters: mean number of alleles per locus (A), mean number of observed alleles (na), effective number of alleles (ne), Nei's genetic diversity indices (*He*); as long as we sampled parasitized trees we used Graphpad Prima 7.0 to perform a chi square test and detect if there were differences in heterozygosis according to the categories considered in sampling. The number of polymorphic alleles (LP), percentage of polymorphic alleles (%LP), population structure fixation indices (G_{ST}), indicators of heterozygosis (H_T: total genetic diversity of the locus, H_S: genetic diversity within populations), gene flow (Nm) and Nei's genetic distance (D) between the four categories we obtained BB, DM, BM, and HT [35].

3. Results

We obtained electrophoretic patterns with high reproducibility and clear band resolution. UBC-254: 11 bands, ASSR-15: 7 bands, ASSR-29: 11 bands, Cox3in: 8 bands and 10FF/RR: 7

	ncDNA				mtDNA				cpDNA			
	BB	DM	BM	HT	BB	DM	BM	HT	BB	DM	BM	HT
n	35	75	35	35	75	35	35	35	75	35	35	34
na	1.9 ± 0.3620	2	1.9 ± 0.3620	1.7 ± 0.4237	1.4 ± 0.5175	1.8 ± 0.4629	1.6 ± 0.5175	1.5 ± 0.5345	1.4 ± 0.5345	2	1.4 ± 0.5345	1.6 ± 0.5345
ne	1.3 ± 0.2964	1.3 ± 0.3067	1.7 ± 0.2872	1.3 ± 0.2666	1.0 ± 0.1433	1.1 ± 0.1859	1.1 ± 0.2395	1.0 ± 0.1047	1.0 ± 0.0395	1.1 ± 0.2033	1.0 ± 0.1056	1.1 ± 0.1661
He	0.1924 ± 0.1648	0.1993 ± 0.1586	0.1877 ± 0.1526	0.1831 ± 0.1552	0.0539 ± 0.1016	0.0810 ± 0.1196	0.0872 ± 0.1382	0.0582 ± 0.0828	0.0295 ± 0.0368	0.1267 ± 0.1260	0.0592 ± 0.0857	ND
% LP	87.5	87.5	100	75.86	23.10	75.86	86.21	33.33	55.56	33.33	55.56	ND

Table 2. Genetic variability of *Pinus hartwegii* in attacked (by bark beetle and dwarf mistletoe) and nonattacked trees for nuclear DNA (ncDNA), mitochondrial DNA (mtDNA), and chloroplast DNA (cpDNA); bark beetle attacked tree (BB), dwarf mistletoe attacked tree (DM), bark beetle and dwarf mistletoe attacked tree (BM) and healthy tree (HT); sample size (n), number of alleles per locus, effective number of alleles per locus (ne), Nei's genetic diversity, percentage of polymorphic loci (%LP), (ND) no data.

bands, combined in 139 band patterns, ranging from 200 to 2000 bp; additionally there were bands only present in HT trees.

3.1. Genetic variability

The number of alleles per locus (na) and the number effective alleles (ne) ranged from 1.3 to 2.0 indicating that the number of alleles transferred from one generation to the next is low; these low values of *na* and *ne* consequently act on heterozygosis (Nei's genetic diversity: *He*), which also was low compared with other *Pinus* species (**Table 2**).

Among categories (BB, DM, BM and HT) we observed, in some cases, a tendency of BB, DM and BM to present higher values of *He*, for example with ncDNA BB and DM categories showed *He* values 0.1924 and 0.1993, respectably, higher than in HT trees which was 0.1831, with cpDNA DM trees presented high *He* compared with HT (**Table 2**). According to Chi square tests, the distribution of *He* showed statistically significant differences between the BB, DM, BM and HT categories (**Table 3**).

3.2. Population structure

Estimated population structure based on G_{ST} (fixation index) was very low probably due to high levels of gene flow (**Table 4**). The rates of gene flow (Nm) derived from G_{ST} were very

Marker	X^2	P
ASSR-15	1.0614	0.0001
ASSR-29	1.8906	0.0001
UBC254	0.8028	0.0001
10FF/RR	0.8732	0.0001
COX3IN	0.5732	0.0001

Table 3. Comparison of genetic variability between groups (BB, DM, BM and HT) in *Pinus hartwegii*.

	ncDNA				mtDNA				cpDNA			
	Mean	SD	MAX	MIN	Mean	SD	MAX	MIN	Mean	SD	MAX	MIN
G_{ST}	0.0321	39.8316	0.1404	0.0006	0.0170	0.0065	0.0230	0.0049	0.0475	0.0312	0.0961	0.0096
Nm	15.0849	10.8853	55.5996	3.0604	28.9288	14.9747	59.5707	21.2152	10.0279	17.0230	51.3333	4.7046
H_T	0.1946	0.0203	0.4906	0.0302	0.0713	0.0122	0.3294	0.0064	0.0832	0.0064	0.2076	0.0064
H_S	0.1884	0.0194	0.4850	0.0120	0.0700	0.0116	0.3218	0.0063	0.0792	0.0058	0.1955	0.0063

Table 4. Summary of *Pinus hartwegii* population differentiation (G_{ST}), genic flow (Nm), total genetic diversity of the locus (H_T), average genetic diversity (H_S), for nuclear (ncDNA), mitochondrial (ncDNA) and chloroplast DNA markers (cpDNA).

		HT	BB	DM	DM
ncDNA	HT	_			
	BB	0.006	_		
	DM	0.014	0.007	_	
	BM	0.012	0.004	0.007	_
mtDNA	HT	_			
	BB	0.002	_		
	DM	0.010	0.001	_	
	BM	0.004	0.001	0.003	_
cpDNA	HT	_			
	BB	0.014	_		
	DM	0.003	0.013	_	
	BM	0.005	0.006	0.001	_

Table 5. Matrix for genetic distance values for bark beetle attacked tree (BB), dwarf mistletoe attacked tree (DM), bark beetle and dwarf mistletoe attacked tree (BM) and healthy tree (HT) in *Pinus hartwegii* samples, evaluated with nuclear (ncDNA), mitochondrial (mtDNA) and chloroplast (cpDNA) markers.

high with all markers, averaging 15.1 migrants per generation (**Table 4**). Based on these results (low-population differentiation/high gene flow), we assume that inbreeding rates are low.

Genetic distances ranged from 0.003 between DM/BM with mtDNA and DNA cp markers to 0.014 between HT/DM and HT/BB with ncDNA and cpDNA markers; for all markers parasitized trees had the highest genetic distances between nonattacked trees (HT) and attacked trees (BB, DM and BM) (**Table 5**).

4. Discussion

We found low levels of genetic variability in mountain pine, but also that parasitized trees in some cases had highest levels of heterozygosity; this is not rare if we take into account that plants are subject to various abiotic and biotic stresses, especially those long-lived species like conifers; thus, slow-growing plants will invest heavily in defenses against parasites because of high cost of replacing tissue [9, 36]; substantial variation in susceptibility, damage and resistance are well documented in natural plant populations [21, 37–39].

4.1. Genetic variability

Levels of heterozygosis in mountain pine were in general low, but in spite of this with ncDNA, we observed a tendency of parasitized trees (BB and DM) to have more genetic variability and this is relevant because in a population stressed by an increment of parasites, high genetic diversity individuals will have more chances to adapt themselves to changes in their environment and attack by parasites and pathogens.

Some theories propose that dwarf mistletoes performance is regulated by physiological (genetic) condition of the host and that infections are greater in sites with high stress, also long-term contact and evolutionary history between specific plant and parasites are expected to increase plant defenses, tree species have been co-evolving with mistletoes for 25 million years, so high genetic variability of parasitized organisms is an insight of co-evolution system [16, 40, 41]. The genetics of host resistance due to a co-evolutionary linkage has been reported in some pine species, such as *P. edulis*, *P. lawsonii*, *P. montezumae* and *Fagus sylvatica* with their associated bark beetles [9, 42], and for *P. ponderosa* and Douglas-fir (*Pseudotsuga menziesii*) to dwarf mistletoes [37, 40, 43, 44].

In PFFANT, pine forests are under high abiotic stress like the presence of human settlements and logging which in turn could be causing biotic stress situations like incrementing incidence of parasites (e.g., because of the loss of trees, bark beetles move to sites where there were no infections and the infection spreads in trees which are predicted to have low defenses). The location of populations parasitized and with high genetic variability is a valuable data to be taken into account for conservation science.

P. hartwegii form dense forest stands in the PFFANT, and their associated parasites are also present with a patchy distribution. They are not present in all populations, indeed there are healthy trees in populations highly parasitized (also called "scape trees") it could suggest that the parasites of mountain pine select their host based on genetic cues. There is a hypothesis called "gene by gene coevolution" which pose that host-parasite relationship had been kept polymorphisms in plants, these has been recognized in the existence of two cases: host-specificity, and variation in host preference [40, 45], we suggest that in view of the levels of *He* in parasitized trees, it is possible that a gene-by-gene coevolution case is present in the interaction of mountain pine with its parasites, although there is a need for more studies to probe it. It appears that we are facing a coevolution dynamics, where trees with higher genetic diversity are attacked due to its genes; parasites track specific host genotypes under natural conditions [46–48] not only to phenotypic characteristics. In other words, plant genetic diversity affects enemies and mutualists [49].

4.2. Population structure

Organelle genomes provide information about the relative capacity of dispersal of males and females; mtDNA is maternally inherited, and cpDNA is paternally inherited in pines, this shapes gene flow and genetic diversity within and among populations in a particular way [22, 50, 51], we found low levels of $G_{ST,}$ in comparison with some other pine species at organelle genomes [24] and registered high levels of gene flow, these data could suggest that mountain pine subpopulations studied at PFFANT are poorly differentiated which in turn could mean that the population as a whole must be managed and conserved as genetically unique; also we registered the highest with mtDNA on this we can infer that *P. hartwegii*'s most important means of gene flow seems to be the seed (**Table 4**), but it is also probably due to the reforestation plans performed at PFFANT by government institutions, which in most of the cases is performed with pines from off-site seed carrying foreign DNA.

High levels of deforestation have been reported at PFFANT, mainly due to illegal logging leading to forest fragmentation, which in turn has devastating effects on forest trees, taking off the best conformed trees, which reduce levels of gene flow and allele diversity, promoting inbreeding and genetic drift [52]. Since conifers are wind pollinated and long lived organisms, effects of inbreeding and or genetic drift are not drastic yet (in view of our results) and the levels of genetic variability reported here may only reflect *standing genetic variation*, when a population passes through selective pressures it adapts mainly from standing genetic variation [53, 54].

Genetic distances differed according to marker type; there has been reported high polymorphism in *Dendroctonus ponderosa* related to geographic region [55]. Our results may be explained by differences in land use in which trees were sampled, which could impose different selection pressures in parasites and hosts in response to biogeographic discontinuities [42, 56]. We observed, for example that dwarf mistletoe infections response to altitude, at least *Arceuthobium vaginatum* and *A. globossum* (parasites of *P. hartwegii*) disappear above 3600 masl (personal observation at FFPANT). The lowest genetic distances we found were between DM and BB trees with the tree markers (**Table 4**), and these trees are genetically similar and have high levels of polymorphism, the differences between HT trees and DM, BB trees may consist in those which parasitized trees possess and make parasites choose them [40, 45], while markers we used are not specifically related with defense and resistance pathways in plants, we report a reliable difference which can serve as a first insight to conduce more studies on genetic resistance of conifers to bark beetles and dwarf mistletoes.

4.3. Conservation implications

The presence of probably immune trees (HT) and resistant trees (BM) must be taken into account to conserve mountain pine forests of the FFPANT, specially confronting an increase of infected areas by bark beetles and dwarf mistletoes; BM are trees whose seeds could be used in reforestation programs, especially in areas where infections are growing, and HT seeds could be used to create barriers which impede the spread of outbreaks.

Pinus forests of the FFPANT are home and a potential distribution area of many species of plants and animals some of which also have high levels of genetic variability which in turn makes FFPANT an important place to conservation of species [4, 57–60]. These forests provide water to urban areas, hence the importance of preserving this natural area should not be dismissed; there is an urgent need to protect these forests, as long as forest fragmentation is increasing.

It has been reported that there are differences in susceptibility to parasites attack among tree species and even among individuals within a species; when different species interact, selective changes may occur as a result of the interaction; models of host–parasite interactions support the idea that variation in host resistance is, at least partially, genetic and assume the presence of genetic variation [23, 46, 48]. In conifers, some theories have been proposed to explain this variability, including variations in terpene or phenolic content [61], differences in constitutive defenses [62, 63] and differences in growing conditions [64, 65]. Population genetics theory predicts that under many selective regimes, fitness will increase the number of heterozygous loci [23]; many authors have reported that genetic variability of trees affected by parasites, in terms of heterozygosity, they report that heterozygous individuals were resistant and suggest that heterozygosity and plant resistance are positively correlated [61, 66–69].

More studies are needed in order to reaffirm the results reported here, may be with other kinds of molecular markers, and to continue the improvement of management and preservation of the FFPANT. Forests worldwide need to be protected in face of an imminent climate change.

Acknowledgements

The authors wish to thank Consejo Nacional de Ciencia y Tecnología (CONACyT) and the Universidad Autónoma del Estado de México (UAEMéx) for financial support. Also, they thank the CIEAF students who helped in the laboratory and fieldwork.

Author details

Rosa Laura Heredia-Bobadilla[1], Guadalupe Gutiérrez-González[1],
Amaury-M. Arzate-Fernández[1]* and Sergio Franco-Maass[2]

*Address all correspondence to: amaury1963@yahoo.com.mx

1 Centro de Investigación y Estudios Avanzados en Fitomejoramiento, Universidad Autónoma del Estado de México, Toluca, Mexico

2 Instituto de Ciencias Agropecuarias y Rurales, Universidad Autónoma del Estado de México, Toluca, Mexico

References

[1] Candeau-Dufat R, Franco-Maass S. Aplicación de un sistema de información geográfica para determinar la relación entre la dinámica y condiciones de vida de la población de las localidades del parque nacional nevado de Toluca y la generación de impactos ambientales en sus ecosistemas. Investigaciones Geográficas Boletín del Instituto de Geografía UNAM. 2000;**62**:44-68

[2] Heredia-Bobadilla RL, Gutiérrez-González G, Franco-Maass SF, Arzate-Fernández AM. Genetic variability of sacred fir (*Abies religiosa*) in the Nevado de Toluca National Park. International Journal of Biodiversity and Conservation. 2012;**4**(3):130-136. DOI: 10.5897/IJBC11.187

[3] Arriola-Padilla V, Estrada-Martínez E, Ortega-Rubio A, Pérez-Miranda R, Gijón-Hernández AR. Deterioro en Áreas Naturales Protegidas del centro de México y del Eje Neovolcánico Transversal. Investigación y Ciencia. 2014;**22**(60):37-49

[4] Villers-Ruiz L, García del Valle L, López-Blanco J. Evaluación de los bosques templados en México: una aplicación en el Parque Nacional Nevado de Toluca. Investigaciones Geográficas Boletín del Instituto de Geografía UNAM. 1998;**36**:7-19

[5] CONANP. Área de Protección de Flora y Fauna Nevado de Toluca [Internet]. 2013. Available from: http://nevadodetoluca.conanp.gob.mx [Accessed: 01-06-2017]

[6] SEMARNAT. Acuerdo por el que se da a conocer el Resumen del Programa de Manejo del Área Natural Protegida con categoría de Área de Protección de Flora y Fauna Nevado de Toluca. Available from: http://www.conanp.gob.mx/que_hacemos/pdf/programas_manejo/RESUMEN_NEVADO_TOLUCA.pdf, [Accessed: 22-11-2016]

[7] Raffa KF, Aukema BH, Bentz BJ, Carroll AL, Hicke JA, Turner MG, Romme WH. Cross-scale drivers of natural disturbances prone to anthropogenic amplification: The dynamics of bark beetle eruptions. Bioscience. 2008;**58**(6):501-517. DOI: 10.1641/B580607

[8] Ramírez-Dávila JF, Porcayo-Camargo E. Estudio de la distribución espacial del muérdago enano (*Arceuthobium* sp.) en el Nevado de Toluca, México, utilizando el método del SADIE. Madera y Bosques. 2009;**15**(2):93-111

[9] Bentz BJ, Hood SMM, Hansen EM, Vandygriff JC, Mock KE. Defense traits in the long – lived Great Basin bristlecone pine and resistance to the native herbivore mountain pine beetle. New Phytologist. 2017;**213**:611-624. DOI: 10.1111/nph.14191

[10] Keane PJ, Kile GA, Podger FD, Brown BN. Diseases and Pathogens of Eucalypts. Australia: La Trobe University Bundoora; 2000. 565 pp

[11] Sáenz C, Snively AE, Linding R. Conservation and restoration of pine forest genetic resources in Mexico. Silvae Genetica. 2003;**52**:233-237

[12] Shaw DC, Chen J, Freeman EA, Braun DM. Spatial and population characteristics of dwarf mistletoe infected trees in an old-growth Douglas-fir-western hemlock forest. Canadian Journal of Forest Research. 2005;**35**:990-1001. DOI: 10.1139/x05-022

[13] FAO. Global Forest Resources Assessment 2015 [Internet]. 2015. Available from: http://www.fao.org/3/a-i4793e.pdf [Accessed: 05-01-2017]

[14] Cibrián-Tovar D, Tulio MJ, Campos BR, Yates HO, Flores L. Insectos Forestales de México. México: Universidad Autónoma Chapingo; 1995. 453 p

[15] Cibrián-Tovar D, Alvarado-Rosales D, García-Díaz SE. Enfermedades Forestales en México. México: Universidad Autónoma Chapingo; 2007. 587 p

[16] Scharpf R, Parmeter JR Jr. Population Dynamics of Dwarf Mistletoe on Young True Firs in the Central Sierra Nevada. California: United States Department of Agriculture; 1982. 9 p. DOI: 10.2737/PSW-RP-161

[17] Way DA. Parasitic plants and forests: A climate change perspective. Tree Physiology. 2011; **31**(1):1-2. DOI: 10.1093/treephys/tpq113

[18] Klutsch JG, Beam RD, Jacobi WR, Negrón JF. Bark beetles and dwarf mistletoe interact to alter downed woody material, canopy structure, and stand characteristics in northern Colorado ponderosa pine. Forest Ecology and Management. 2014;**315**:63-71. DOI: 10.1016/j.foreco.2013.12.024

[19] González-Martínez S, Krutovsky KV, Neale DB. Forest-tree population genomics and adaptive evolution. New Phytologist. 2006;**170**:227-238. DOI: 10.1111/j.1469-8137.2006.01686.x

[20] Neale DB, Kremer A. Forest tree genomics, growing resources and applications. Nature Reviews Genetics. 2011;**12**:111-122. DOI: 10.1038/nrg2931

[21] Burdon RD. Genetic diversity and disease resistance: Some considerations for research, breeding, and deployment. Canadian Journal of Forest Research. 2001;**31**:596-606. DOI: 10.1139/x00-136

[22] Petit RJ, Duminil J, Fineschi S, Hampe A, Salvini D, Vendramin GG. Comparative organization of chloroplast, mitochondrial and nuclear diversity in plant populations. Molecular Ecology. 2005;**14**:689-701. DOI: 10.1111/j.1365-294X.2004.02410.x

[23] Hedrick PW. Genetics of Populations. Canada: Jones and Barlett Learning; 2011. 675 p

[24] Floran V, Sestras RE, García-Gil MR. Organelle genetic diversity and phylogeography of scots pine (*Pinus sylvestris L.*). Notulae Botanicae Horti Agrobotanici Cluj-Napoca. 2011; **39**(1):317-322. DOI: 10.15835/nbha3916103

[25] Arriaga L, Espinoza JM, Aguilar C, Martínez E, Gómez L, Loa E. Regiones terrestres prioritarias de México. México: Comisión Nacional para el Conocimiento y uso de la Biodiversidad; 2000

[26] Zhou Z, Miwa M, Hogetsu T. Analysis of genetic structure of a *Suillus grevillei* population in a *Larix kaempferi* stand by polymorphism of inter-simple sequence repeat ISSR. New Phytologist. 1999;**144**:55-63. DOI: 10.1046/j.1469-8137.1999.00504.x

[27] Yamagishi M, Abe H, Nakano M, Nakatsuka A. PCR-based molecular markers in asiatic hybrid lily. Scientia Horticulturae. 2001;**96**:225-234. DOI: 10.1016/S0304-4238(02)00095-X

[28] Wu J, Krutovskii KV, Strauss SH. Nuclear DNA diversity, population differentiation, and phylogenetic relationships in the California closed-cone pines based on RAPD and allozyme markers. Genome. 1999;**42**:893-908. DOI: 10.1139/g98-171

[29] Yamagishi M, Abe H, Nakano M, Nakatsuka A. PCR-Based Molecular Markers in Asiatic Hybrid Lily. Scientia Horticulturae. 2002;**96**:225-234

[30] Wen-Qing Q, Yang HJ, Xue YB, Hu SY. Inheritance of chloroplast and mitochondrial DNA in Chinese fir *Cunninghamia lanceolata*. Acta Botánica Sinica. 1999;**41**(7):695-699

[31] Marshall HD, Newton C, Rotland K. Chloroplast phylogeography and evolution of highly polymorphic microsatellites in lodgepole pine (*Pinus contorta*). Theoretical Applied Genetics. 2002;**104**:367-378. DOI: 10.1007/s001220100687

[32] Yeh FC, Rongcai R, Boyle T, Ye Z, Mao-Xiyan J. Pop Gene Version 1.3. Canada: University of Alberta; 1999

[33] Peakall R, Smouse PE. GENALEX 6: Genetic analysis in Excel Population genetic software for teaching and research. Molecular Ecology Notes. 2006;**6**:288-295

[34] Miller MP. Tools for Population Genetic Analyses (TFPGA). USA: Northern Arizona University; 1997

[35] Nei M. Estimation of average heterozygosity and genetic distance from a small number of individuals. Genetics. 1978;**89**(3):583-590

[36] Polyakova GG, Pashenova NV, Polyakov VI, Zrazhevskaya GK. Induction of conifer immune responses by phytopathogenic fungus metabolites. Russian Journal of Plant Physiology. 2008;**55**(4):496-502. DOI: 10.1134/S1021443708040109

[37] Roth LF. Resistance of ponderosa pine to dwarf mistletoe. Silvae Genetica. 1974;**23**:116-119. DOI: 10.2737/PSW-RP-208

[38] Roth LF. Genetic control of dwarf mistletoe. In: Scharpf RF, Parmeter JR Jr, editors. Symposium Proceedings: Dwarf Mistletoe Control through Forest Management; 11–13 April 1978; Berkeley, CA. Gen. Tech. Rep. PSW-31. Berkeley, CA: U.S. Department of Agriculture, Forest Service, Pacific Southwest Forest Range and Experiment Station; 1978. pp. 69-72

[39] Yanchuk AD, Murphy JC, Wallin KF. Evaluation of genetic variation of attack and resistance in lodgepole pine in the early stages of a mountain pine beetle outbreak. Tree Genetics and Genomes. 2008;**4**:171-180. DOI: 10.1007/s11295-007-0098-9

[40] Shamoun SF, DeWald LE. Management strategies for dwarf mistletoes: Biological, chemical, and genetic approaches. USDA Forest Service General Technical Report; 2002. pp. 75-82

[41] Bickford CP, Kolb TE, Geils BW. Host physiological condition regulates parasitic plant performance: *Arceuthobium vaginatum* subsp. *cryptopodum* on *Pinus ponderosa*. Oecologia. 2005;**146**:179-189. DOI: 10.1007/s00442-005-0215-0

[42] Krabel D, Petercord R. Genetic diversity and bark physiology of the europan beech (*Fagus sylvatica*): A coevolutionary relationship with the beach scale (*Cryptococcus fagisuga*). Tree Physiology. 2000;**20**:485-491. DOI: 10.1093/treephys/20.7.485

[43] Scharpf RF. Host resistance to dwarf mistletoes. In: Hawksworth FG; Scharpf RF, editors. Biology of Dwarf Mistletoes, Symposium Proceedings; 8 August 1984; Fort Collins, CO. Gen. Tech. Rep. RM-111. Fort Collins, CO: U.S. Department of Agriculture, Forest Service, Rocky Mountain Forest and Range Experiment Station. pp. 70-76

[44] Scharpf RF. Resistance of Jeffrey pine to dwarf mistletoe. In: Weber HC, Forstreuter W, editors. Proceedings of the 4th International Symposium on Parasitic Flowering Plants; 2–7 August 1987. pp. 745-753

[45] Thompson JN, Burdon JJ. Gene-for-gene coevolution between plants and parasites. Nature. 1992;**360**:121-125. DOI: 10.1038/360121a0

[46] Sorci G, Møller AP, Boulinier T. Genetics of host-parasite interactions. Science. 1997;**12**(5): 196-200

[47] Carius HJ, Little TJ, Ebert D. Genetic variation in a host-parasite association: Potential for coevolution and frequency-dependent selection. Evolution. 2001;**55**(6):1136-1145. DOI: 10.1111/j.0014-3820.2001.tb00633.x

[48] Koskela T, Salonen V, Mutikainen P. Interaction of a host plant and its holoparasite: Effects of previous selection by the parasite. Journal of Evolutionary Biology. 2001;**14**:910-917. DOI: 10.1046/j.1420-9101.2001.00352.x

[49] Moreira X, Mooney KA. Influence of plant genetic diversity on interactions between higher trophic levels. Biology Letters. 2013;**9**:20130133. DOI: 10.1098/rsbl.2013.0133

[50] Neale DB, Sederoff RR. Inheritance and evolution of conifer organelle genomes. In: Hanover JW, Keathley DE, Wilson CM, Kuny G, editors. Genetic Manipulation of Woody Plants. Basic Life Sciences. Vol. 44. USA: Springer; 1998

[51] Hipkins VD, Krutovskii KV, Strauss SH. Organelle genomes in conifers: Structure, evolution, and diversity. Forest Genetics. 1995;**1**(4):179-189

[52] Jump AS, Peñuelas J. Genetic effects of chronic habitat fragmentation in a wind-pollinated tree. Proceedings of the National Academy of Sciences. 2006;**103**(21):8096-8100. DOI: 10.1073/pnas.0510127103

[53] Hermisson J, Pennings PS. Soft sweeps: Molecular population genetics of adaptation from standing genetic variation. Genetics. 2005;**169**:2335-2352. DOI: 10.1534/genetics.104.036947

[54] Barret RDH, Schluter D. Adaptation from standing genetic variation. Trends in Ecology and Evolution. 2007;**23**(1):38-44. DOI: 10.1016/j.tree.2007.09.008

[55] Stock MW, Amman GD, Bentz JB. Isozyme studies of barck beetle population genetics and systematics. USDA Forest Service General Technical Report. 1992;**138**:7-9

[56] Leonardi S, Piovani P, Scalfi M, Piotti A, Giannini R, Menozi P. Effect of habitat fragmentation on the genetic diversity and structure of peripheral populations of beech in central Italy. Journal of Heredity. 2012;**103**(3):408-417. DOI: 10.1093/jhered/ess004

[57] Sánchez-Jasso J, Aguilar-Miguel MX, Medina-Castro JP, Georgina M. Riqueza específica de vertebrados en un bosque reforestado del Parque Nacional Nevado de Toluca. Revista Mexicana de Biodiversidad. 2013;**84**:360-373

[58] Medina JP, Salgado-Miranda C, García-Cornejo M, Soriano-Vargas E. Coccidia in passerines from the Nevado de Toluca National Park, México. Acta Parasitologica. 2015;**60**(1): 173-174. DOI: 10.1515/ap-2015-0024

[59] Heredia-Bobadilla RL, Monroy-Vilchis O, Zarco-González MM, Martínez-Gómez D, Mendoza-Martínez G, Sunny A. Genetic structure and diversity in an isolated population of an endemic mole salamander (*Ambystoma rivulare* Taylor, 1940) of Central México. Genetica. 2016;**144**(6):689-698. DOI: 10.1007/s10709-016-9935-9

[60] Sunny A, González-Fernández A, D'Addario M. Potential distribution of the endemic imbricate alligator lizard (*Barisia imbricate imbricata*) in highlands of central México. Amphibia-Reptilia. 2017;**38**:225-231. DOI: 10.1163/15685381-00003092

[61] Liu JJ, Ekramoddoullah AKM, Hunt RS, Zamani A. Identification and characterization of random amplified polymorphic DNA markers linked to a major gene (Cr_2) for resistance to *Cronartium ribicola* in *Pinus monticola*. Genetics and Resistance. 2006;**96**(4):395-399. DOI: 10.1094/PHYTO-96-0395

[62] Eyles A, Bonello P, Ganley R, Mohammed C. Induced resistance to pests and pathogens in trees. New Phytologist. 2009;**185**:893-908. DOI: 10.1111/j.1469-8137.2009.03127.x

[63] Pastorino RM, Corley J. Bases genéticas de la resistencia de los árboles a las plagas, Serie Técnica: Manejo Integrado de Plagas Forestales; 2009. 15 p

[64] Beaulieu J, Plourde A, Daoust G, Lamontagne L. Genetic variation in juvenile growth of *Pinus strobus* in replicated Quebec provenance-progeny tests. Forest Genetics. 1996;**3**(2): 103-112

[65] King JN, Yanchuk AD, Kiss GK, Alfaro RI. Genetic and phenotypic relationships between weevil (*Pissodes strobi*) resistance and height growth in spruce populations of British Columbia. Canadian Journal of Forest Research. 1997;**27**(5):732-739. DOI: 10.1139/x97-009

[66] Benet H, Guries RP, Boury S, Smalley EB. Identification of RAPD markers linked to a black leaf spot resistance gene in Chinese elm. Theoretical and Applied Genetics. 1995;**90**:1068-1073. DOI: 10.1007/BF00222923

[67] Devey ME, Delfino-Mix A, Kinloch BB Jr, Neale DB. Random amplified polymorphic DNA markers tightly linked to a gene for resistance to White pine blister rust in sugar pine. Proceedings of the National Academy of Sciences. 1995;**92**(6):2066-2070

[68] Kondo T, Terada K, Hayashi E, Kuramoto N, Okamura M, Kawasaki H. RAPD markers linked to a gene for resistance to pine needle gall midge in Japanese black pine (*Pinus thunbergii*). Theoretical and Applied Genetics. 2000;**100**(3):391-395. DOI: 10.1007/s001220 050051

[69] Mopper S, Mitton JB, Whitham TB, Cobb NS, Christensen KM. Genetic differentiation and heterozygosity in pinyon pine associated with resistance to herbivory and environmental stress. Evolution. 1991;**45**(4):898-999. DOI: 10.1111/j.1558-5646.1991.tb04365.x